The Social Life of the Lion

The Social Life of the Lion

A study of the behaviour of wild lions (Panthera leo massaica [Newmann]) in the Nairobi National Park, Kenya

Judith A. Rudnai

WASHINGTON SQUARE EAST, PUBLISHERS
Wallingford, Pennsylvania 1973

Published in U.K. by
MTP
Medical and Technical Publishing Co Ltd
St. Leonards House, St. Leonardgate
Lancaster England

Published in USA by Washington Square
East, Publishers
Wallingford, Pennsylvania, USA

ISBN-13: 978-94-011-7142-7   e-ISBN-13: 978-94-011-7140-3
DOI: 10.1007/978-94-011-7140-3

Contents

I INTRODUCTION   I
  The Habitat   I
  History and physical features of Nairobi National Park   I
  The animal population   3
  Aims and scope of this study   3
  Methods and procedures   4
  Identification method   5
  Ageing   6
  The lion. Description and distribution   6

II POPULATION STRUCTURE AND RANGES   11
  Population size, composition and density   11
  The prides: composition and changes throughout the year   12
  Home ranges   15
  Day range utilization   20
  Habitat preferences   21

III ACTIVITY PATTERNS   24
  Daily activity cycles   24
  Daytime activities   24
  Activity pattern of females   25
  Activity pattern of male   27
  Cubs' activity pattern   28
  Resting positions; degrees of alertness   29
  Night activities   30
  Distances covered by night   30
  Range utilization by night   33

IV INDIVIDUAL ACTIVITIES   34
  Locomotion   34
  Speed   34
  Pattern of walking   35
  Tree climbing   37
  Health and self-care   37
    Injuries, diseases and parasites   37
    Body functions   38
    Self-care and comfort activities   39

V SOCIAL BEHAVIOUR   44
  Group life; synchronized activities   44
  Amicable behaviour   46
    Mutual grooming   46
    Greeting (headrubbing)   48
  Agonistic behaviour   49
    Threat and intimidation displays   49
    Defence; submission   51
  Intragroup relationships; social bonds, individual associations,
    rank order, leadership   52
  Communication: vocalization and marking   55
  Intergroup relationships; meeting of prides   60

VI REPRODUCTION AND DEVELOPMENT   62
  Sexual behaviour; courtship and mating   62
  Gestation period, litter size, cub mortality   67
  Physical development and care of young   69
  Feeding and grooming of cubs   70
  Protection of cubs   71
  Vocal communication between mother and cub   72
  Play and the development of behaviour patterns and social
    interactions   72
  Development of sexual dimorphism and independence   75

VII PREDATION   77
  Predation patterns   77
    The prey animals; numbers, density   77
    Prey-predator ratios   78
    Prey selection (based on carcase analysis)   80
    Habitat as a factor in prey selection   84
  Hunting methods and feeding behaviour   85
    Searching   85
    Preference ratio based on stalks   87
    Attacking and killing   89
    Feeding   91
    Consumption; kill frequency   94
    Drinking   96
    Hunting success   97

VIII REACTION TO ENVIRONMENT   98
  Reaction to humans and vehicles   98
  Interaction with other species   99
  Response to weather   102

IX DISCUSSION AND EPILOGUE   103

APPENDICES   107

REFERENCES   114

ACKNOWLEDGEMENTS   118

INDEX   119

1 Introduction

THE HABITAT

Nairobi National Park lies to the south–southeast of the capital of Kenya, Nairobi, where the Athi Plains meet the Eastern escarpment of the Rift Valley. These plains form part of the semi-arid highland plateau lying between the coast and the Rift Valley.

Both the city itself, and the Park bordering it, are a meeting place of two generally distinct types of landscape and climate. While to the east-southeast are semi-arid plains with grasslands and scattered trees, the western-north-western parts are higher, hilly, cooler, more humid and support lush forests.

The combination of latitude—a little over one degree south of the Equator—and altitude—average of 1600 m. (5000–5500 ft.)—combine to give Nairobi a most equable climate where the temperature varies during the year between about 11 degrees and 27 degrees centigrade (mean minimum and mean maximum for eight years). However, considerable changes are usually experienced within each day and a rise from 14 degrees C at 0600 hours to 22 degrees at 1100 hours is not unusual.

Nairobi Park has a unique concentration of wild animals living in their natural habitat less than 10 km. from the centre of a modern city of half a million people. The only interference with the natural course of events in the Park is that normally required for the proper management of a game park, such as maintenance of roads and dams, and, in this particular case, partial fencing towards the city.

HISTORY AND PHYSICAL FEATURES OF NAIROBI NATIONAL PARK

In 1933 Captain Ritchie, game warden for Kenya, proposed that part of the Southern Game Reserve bordering the city of Nairobi be converted into a National Park. By 1939 boundaries were established, but the Park was not officially gazetted until 1945 (Cowie, 1951). It was then gradually developed: roads were built, salt licks established and dams constructed along the waterways to control flooding and provide more permanent water. The only permanent river in the area is the Embakasi or Mbagathi, forming the southern boundary of the Park. The dams contain water in any but the severest dry season and a few isolated pools usually persist even in the otherwise dried-out rivercourses.

It became eventually necessary to prevent the animals from straying into the city, causing hazards on the highways and raiding agricultural areas towards the eastern plains and the western Ngong Hills. At present all but the southern boundary is fenced with strong wire mesh so that, apart from an occasional escapade, animals are free to move only in that direction. The Kitengela Conservation area adjoins the Park from the south and it is from and to that area, forming part of the Athi Plains, that movement of game still occurs. Some species use the Park as a dry season concentration area and disperse to the south in the wet season, but a certain amount of movement occurs continuously irrespective of season.

The Kenya National Parks are at present negotiating to purchase additional land in the Kitengela Conservation unit so as to include some of this wet season dispersal area in the Park, as it is feared that the inevitable increase in settlement along the southern border would eventually confine all the game to within the present boundaries thus seriously threatening the continued existence of the area as a Game Park.

As National Parks go, this is a small unit, covering only 114·8 km², just under 8 km. at its widest point from north to south and stretching for 30 km. from west to east (Fig. I–1).

The largest part of the Park consists of plains sloping up from 1520 m. altitude at the southeastern end to 1700 m. in the west, bisected by numerous river valleys with seasonally flowing rivers, running in a northwest-southeasterly direction in a dendritic pattern. These valleys deepen into rocky gorges at their southern section before their junction with the Embakasi river.

Fig. I-1
(From Foster &
Coe, 1968)

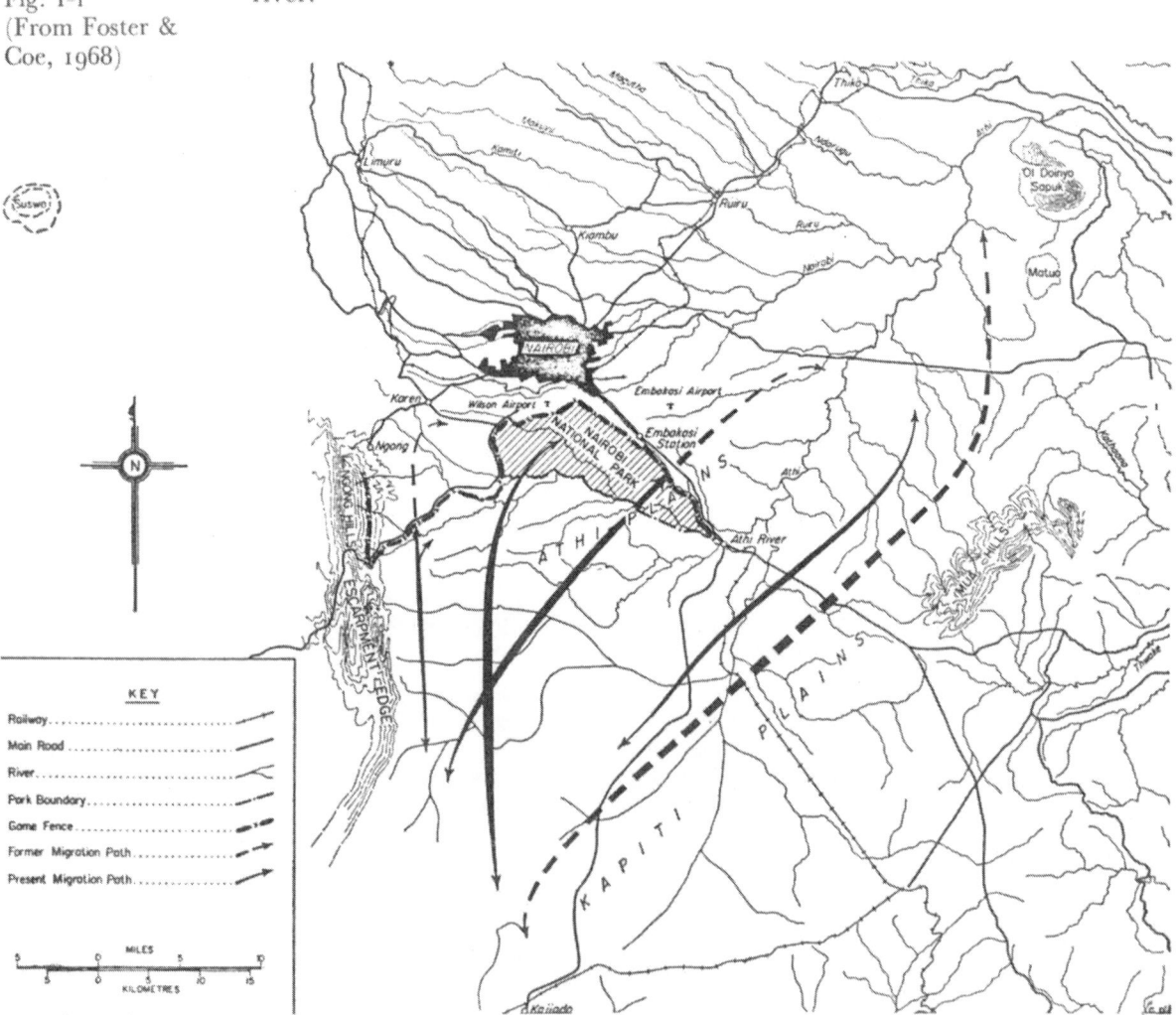

A more elevated area at the western end, between 1700 m. and 1800 m., consists of forest and woodland, ranging from scattered trees and low shrubs to patches of dense growth, with a canopy of 10–15 m., where some of the characteristic trees are olive, cape chestnut and croton.

On the plains there are a number of distinct types of vegetation associated with the changing soil types and the changes in topography. (See Appendix I).

A very characteristic feature of this Park is the extensive area of Acacia scrub; about 40% of Nairobi Park, along the nothern border, is covered with stands of 'whistling thorn' (*Acacia drepanolobium*). These trees, no higher than 1–2 m., form stands of varying density on medium high and high grass, with no other trees or bush associated with them except for a very occasional Balanites. This type of vegetation occurs on flat or gently undulating country on 'black cotton soil', a type of margalithic earth, which becomes extremely sticky when wet, due to impeded drainage. Thus, during the rainy season, some of this area becomes impassable even to four-wheel drive vehicles.

Most of the rest of the plains, except for the river valleys, consists of grassland with scattered bushes and trees, the latter being mostly various species of Acacia.

The valley sides and gorges are covered with coarse grasses, scattered bush and thicket whilst the rivercourses themselves are lined with a narrow strip of riverine forest where the yellow-barked Acacia (*A. xantophloea*), the 'fever tree' of the early explorers, predominates.

A distinct feature of the grasslands and the valleys are the old termitaria conspicuous as slight elevations or mounds up to about 1 m. high with gently sloping sides and supporting a somewhat denser vegetation of herbs and small shrubs.

THE ANIMAL POPULATION

Thanks to the variety of habitats within a relatively small area, Nairobi Park supports an abundance of wildlife representing many different species. About eighty species of mammals have been recorded although some of them are strictly nocturnal, like the genet or the spring hare, and others very secretive and shy, like the bat-eared fox, and thus are hardly ever encountered by visitors.

About fifteen species of herbivores, ranging from the tiny dik-dik (height at shoulder 38 cm.) to buffalo and rhino, are commonly seen; hippo are resident in the Embakasi river and crocodiles live in both the Embakasi river and in some of the dams.

Out of the many species of primates recorded olive baboon and vervet monkey are most often seen.

Among the carnivores all the large cat species are present, lion and cheetah are seen on most days, but leopards only rarely as they inhabit the forest area and some of the riverine vegetation and are more strictly nocturnal and secretive than either lion or cheetah.

Although both side-striped and silver-backed jackals are recorded, only the latter are commonly seen. Hyaenas have become very rare since the early sixties and hunting dogs are only seen a few times a year on their periodic short visits to the Park. (See Appendix II for list of mammals.)

AIMS AND SCOPE OF THIS STUDY

Lions are not only the most important predators in Nairobi Park but by all counts the greatest attraction to visitors. They are usually easily seen and have become habituated to cars so that they can be approached reasonably closely without causing disturbance or forcing them to move away. This fact also facilitates their study. Another factor assisting in the study or evaluation of the feeding habits of this predator is the fact that monthly game censuses have been conducted in Nairobi Park for a number of years so that the approximate number and constitution of the prey population is known: thus changes in prey numbers throughout the years and throughout the seasons can be followed.

Although many observations had been made on lions in general and the Nairobi lions in particular when the project described in this book was undertaken no systematic study on the behaviour, food habits and social system of wild lions had been carried out. The aim of the project was to investigate hunting behaviour, kill frequency, selection of prey as to species, age and sex, and social behaviour of a wild lion population. Any other data on habits and ecology that became available were also recorded.

All information on the ecology and behaviour of any wild species in a National Park, but especially of an animal that is the most important predator as well as the most sought after attraction, is highly desirable and may become of paramount importance whenever control measures of any type, or other basic policy decisions are to be considered. At such times these data must already be on hand as whenever some new management practice becomes necessary the matter is usually of considerable urgency. This study was aimed at providing such basic data on which to base future management policy.

METHODS AND PROCEDURES

The data here presented are based on field observations between January 1968 and July 1969. Periodic trips between January and June 1968 and daily trips from June 1968 to June 1969 were made into the Park in a Land Rover. Occasional trips were again made after June 1969 until the end of that year, and some observations from this period are also included.

On the basis of the lions' position the previous night, a search was made each morning, utilizing spoor on the roads and other signs such as the line of bent grasses in the wake of passing lions. The head ranger, with one or two of his staff of rangers, search the Park every morning routinely to locate the most important tourist attractions such as lions, cheetah and rhino. I followed them on these rounds in my own Land Rover. They had the advantage of superior eyesight and more than one pair of eyes for locating lion spoor, which, in most cases seemed extremely elusive to my vastly inferior eyes. When lions were located, either this pride was kept under observation for the rest of the time spent in the Park or an attempt was made to locate other animals. If the lions were on the move, they were followed until they settled down for the day and were then kept under observation until the closing of the Park, or darkness, at 1900 hours. Alternatively observations were terminated when the last lion disappeared into a thicket or other cover and then were resumed at about 1600 or 1700 hours, at the time activity started again.

Zeiss binoculars of 8 × magnification were used and black and white photographs were taken with a 35 mm. Minolta camera fitted with a 200 mm. telephoto lens. Air temperature was recorded with a whirling hygrometer at hourly intervals; a Rocar stopwatch was used for timing.

The following data were recorded on a daily record sheet: time of entry; time of locating lions; time of leaving lions; date; cloud cover; condition of grass; temperature; position on map; type of cover; activity; composition of pride; pride identification; individual identification. In addition, a running record of the lions' activities was entered in a field note book.

During night observations the pride was contacted at about 1800 hours, then followed and observed by moonlight with binoculars from a distance so as not to cause disturbance to the animals. Records were taken with a Phillips tape recorder and later transcribed. These night observations were attempted for about five nights each month during full moon.

IDENTIFICATION METHOD

For any study of social behaviour it is imperative that individuals be easily and reliably recognized. Recognition by scars and torn ears was tried, but found unsatisfactory as two different animals can easily receive similar scars: moreover, scars proved to be impermanent in many cases. These distinguishing marks were also conspicuous by their absence in younger animals. Thus a reliable, objective recognition method was sought and the system described below was developed. It was found to be satisfactory, as all animals encountered during the study could be distinguished by it.

The method is based on the pattern of dark spots accompanying the vibrissae on the lion's muzzle. The number and position of the spots in the uppermost row of the complete pattern vary from individual to individual, moreover, they differ on the two sides of the same animal's face, giving sufficient variation for identification. For a full description of the method and its application in the field, see Appendix III.

Fig. I-2

Hour of day	Hours of observation
18.00–19.00	142
17.00–18.00	89
16.00–17.00	32·5
15.00–16.00	23·5
14.00–15.00	28
13.00–14.00	39·5
12.00–13.00	69·5
11.00–12.00	103
10.00–11.00	103·5
09.00–10.00	95
08.00–09.00	92·5
07.00–08.00	72·5
Total	890·5

AGEING

To arrive at an approximate age of cubs a combination of factors was considered. (a) As noted by Schenkel (1966) and Schaller (1969) and confirmed during this study, the lioness temporarily leaves her pride for the period of parturition and for the first six to eight weeks after the birth of the litter. Cubs are usually introduced to the pride when they are at least four to six weeks old. (b) The prominence of the nipples during lactation gives a clue to the fact that a lioness has young cubs, even though they are not seen. (c) Dentition of two cubs in captivity was described by David (1962): Incisors and canines were clear of their sheath by the forty-second day and dentition, including molars, was complete by the sixtieth day. All incisors and canines were through by the sixth week in a known-age lion cub reared by Mrs. W. Woodley (pers. com.). (d) The spot pattern on the animals' body was another indication of approximate age. Cubs of about six to eight weeks have dark, solid spots all over their body, developing very shortly into rosettes—similar to the pattern on the leopard's coat—except on the forehead where the spots remain solid. These markings gradually fade away.... in males earlier than in females, ... first disappearing from the back, while remaining quite distinct on the head, belly, sides and inside the upper legs. Next to disappear are the spots on the forehead and sides. Those on the belly and inside upper legs often persist, especially in lionesses, well into adulthood: even females with fully grown young may show some spotting. (Plate 3 a, b, c, d).

The probable date of birth of cubs was determined on the basis of these data, but all ages given for cubs in this report are approximate.

In adult and subadult animals relative age was indicated partly by the presence or otherwise of juvenile spots and partly by the wear of the canines. No attempt was made to obtain exact ages.

THE LION. DESCRIPTION AND DISTRIBUTION

Maybe no other animal has excited man's imagination more than the lion. Since prehistoric times it has been a favourite subject of artists and has played a prominent part in legend and fable. It was considered in turn savage, gentle, blood-thirsty, magnanimous, vicious, peaceful and mild (Guggisberg, 1961). Lions were and still are one of the main attractions which bring travellers to this Continent, they are also the most important predators wherever they occur. Yet, there was no published systematic study of their ecology when this project was undertaken. The most comprehensive book on the subject was C. A. W. Guggisberg's 'Simba' which brought together information gathered at first hand in Nairobi National Park and data found scattered in the literature from ancient times to the recent accounts of travellers and explorers.

(For other studies on the lion see Appendix IV.)

The lion is one of the largest members of the Felidae, exceeded in size only by the tiger. Head and body length recorded by Meinertzhagen (1938) range between 2·4 m. and 2·7 m. for lionesses and 2·6 m. and 2·9 m. for lions. In Rowland Ward's *Records of Big Game* (1962, 1966) both these maxima are exceeded: the figures given are 2·9 m. and 3·2 m. respectively. Weights as quoted by Meinertzhagen are between 122 kg. and 185 kg. for females and between 149 kg. and 191 kg. for males. Comparable figures for tigers are 2·7 m. to 3·7 m. and 227 kg. to 272 kg. for both sexes (Walker, 1968).

Lions and tigers are amongst the largest Carnivora, together with the polar bear, the grizzly bear and the Alaskan brown bear. The latter, the largest living carnivore, may weigh as much as 780 kg., while weights for the grizzly and the polar bear range up to 360 kg. and 720 kg. respectively (Walker, 1968).

Despite a remarkable basic uniformity of body structure in the Felidae, there exists a great variety of size and colour pattern that gave rise to a correspondingly great taxonomic confusion.

Four subfamilies are now generally recognized, after Haltenorth (1953). The subfamily Pantherinae to which the lion belongs, contains two Genera, *Panthera* and *Uncia,* with four species in the former genus: lion, tiger, jaguar and leopard. (See Appendix V for further details of the lion's taxonomy.)

The Pantherinae have been called 'roaring cats' by Pocock in contrast to the 'purring cats' of the other subfamilies. This distinction is based on the structure of the hyoid which is supplied with an elastic ligament in the Pantherinae, giving a freedom of movement to the larynx enabling these cats to roar. This characteristic was already recognized in the lion by Owen in 1835 (Mazak, 1965).

Pantherinae as well as Felinae and Lyncinae, have completely rectractile claws, shielded by a fold of skin when not in use (Plate 4a and b).

The lion resembles all other felids in its lithe, muscular, deep chested body with rounded and shortened head and reduced dentition. The dental formula is:

$$\frac{3 \quad 1 \quad 3 \quad 1}{3 \quad 1 \quad 2 \quad 1}$$

a total of thirty. The canines are elongated, recurved and somewhat longer on the upper than on the lower jaw (Plate 4c). The carnassials are well developed.

The tongue is covered with sharp, pointed, horny, backward slanting papillae and is an excellent tool for scraping meat off bones.

The oval pupils of the Pantherinae contract to a pinpoint in strong light, while the pupils of Felinae contract to a vertical slit. This was one of the characteristics on which separation of the two subfamilies was based.

The colour of the lion is a tawny yellow, almost exactly matching the colour of the dry grassy plains. The colour results from a combination of shorter sandy-yellow hairs, mixed with and overlain to a varying degree by longer, black guardhairs. Thus the overall impression may vary from a pale sandy colour to a fairly dark tan with a blackish cast.

The throat, abdomen and the insides of the legs are creamy white. A prominent black patch on the back of the ears may often be the only conspicuous feature on a lioness lying in the tawny grass.

The juvenile spot pattern was discussed in the section on ageing.

There is a well-marked dark crestlike line in cubs and females, starting from the top of the head and fading away at the shoulder region, formed by dark guard hairs facing backwards and meeting above the vertebral column.

A whorl formed by hair coming from different directions, found in most small cats behind the ear, is on lions usually located further back on the shoulder region. It is clearly visible on Plate 4d. Another peculiarity of hair-growth on the lion is a large patch of reverse hair growing forward, starting with a vortex above the lumbar vertebrae and continuing forward up to about the middorsal region where it forms a small transverse crest when it meets the hair growing in the opposite direction (Plate 4 d). This patch extends laterally about one-third down the side of the body.

One of the best known features of the lion, the mane of the adult male, is extremely variable both as to colour and size. It may vary from pale yellow through light and dark brown or reddish brown to black. As for extent, it may on occasion, be entirely absent, or consist of only a ruff on top of the head and behind the ears. At the other extreme we find the full, heavy growth of hair, framing the face, covering the neck and extending down to the belly and back from the head well over the shoulders. Tufts of hair may also be present on the elbows of the front legs.

In contrast to some of the other Felidae which have a marked tendency towards melanism (Guggisberg, 1961) melanistic forms of lion are extremely rare and only one instance has been recorded of body colouring sufficiently dark to be regarded as melanistic (Mazak, 1964). One reason for this may be that the preferred habitats of the lion are open grassy plains or savannah, where a black animal the size of a lion would be extremely conspicuous, an undesirable trait in a predator that relies largely on concealment in approaching its prey. Both leopard and tiger, on the other hand, are mainly forest dwelling, a habitat where black colouring is not necessarily a disadvantage.

The occasional occurrence of a blackish or definitely black mane, considered a tendency towards melanism by Guggisberg (op. cit.) and Hemmer (1962, cited by Mazak, 1964) is not regarded as such by Mazak (op. cit.) who contends that melanism, a developmental defect, usually affects the whole body, only very rarely parts of the body in isolation. Another factor he mentions is that on young animals the mane is always tawny at first, only in later years does it occasionally turn dark.

Fig. I–3
The Geographical
Distribution of the
Lion
(From Guggisberg,
1961)

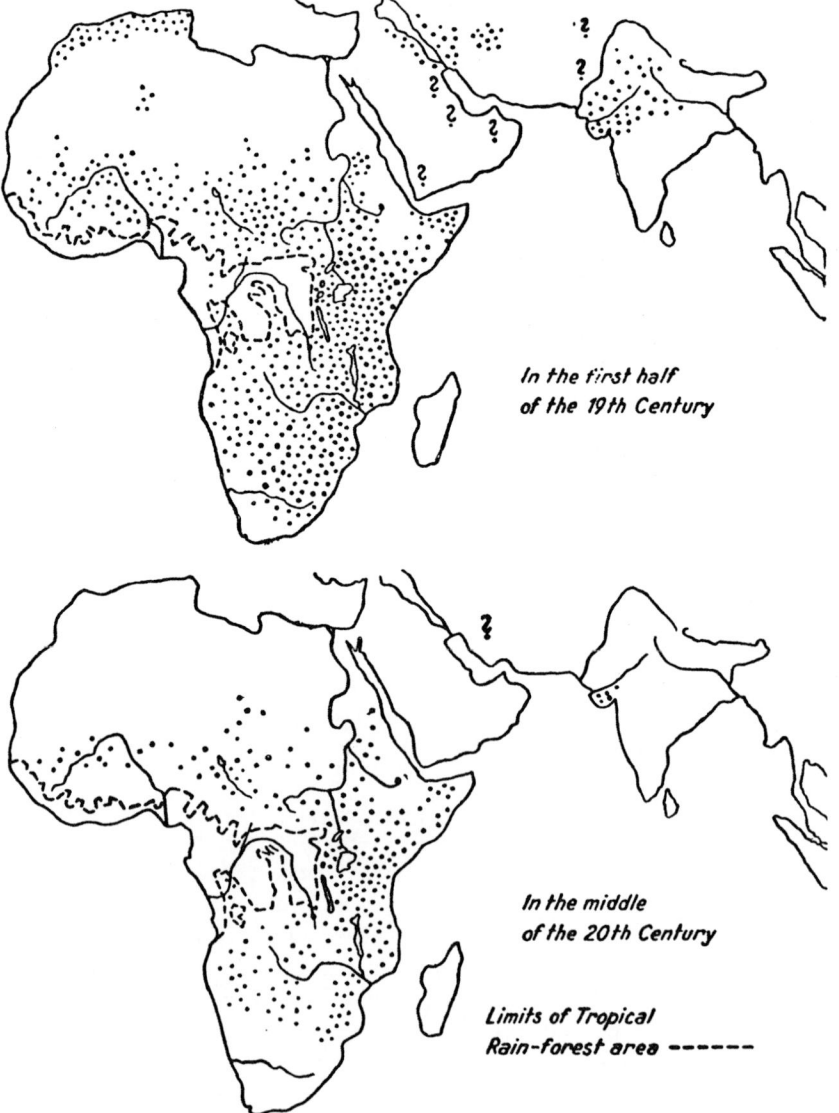

In the first half
of the 19th Century

In the middle
of the 20th Century

Limits of Tropical
Rain-forest area ------

The colour of the mane was formerly used as a basis for subspecific determination but it is not accepted as a criterion for such division by either Mazak (1965) or Guggisberg (op. cit.). However, the position and extent of the mane may be considered as a guide to subspecific differences. One of the characteristics of the now extinct subspecies *P. leo capensis*, or Cape lion, was its extensive mane reaching down to the belly (Guggisberg, op. cit.). Nevertheless, within the one recognized subspecies *P. leo massaica*, great differences may occur in both colour and extent of the mane. Both the large-maned lions well-known from the Serengeti and Ngorongoro areas of Tanzania, and the maneless animals to be found in the Tsavo area of Kenya, made famous by the Tsavo man-eaters at the turn of the century, are considered as belonging to the same subspecies.

(See Appendix V for a list of proposed subspecies.)

The lion differs in some respects from all other felids, including other species of its genus. It is the only cat showing strong gregarious tendencies and the only one to exhibit marked sexual dimorphism. It is also unique in having a horny spur at the tip of its tail, covered by a tuft of black hair. The function of this feature has never been satisfactorily explained.

The evolution of the felids is not well known. During the Pleistocene the Felidae produced some large species: *Felis bebbii* in America and the giant cave lion in the Old World, *Felis spelaeus*. The latter is considered by some a lion, by others, a tiger, still others believe it to have been a distinct species different

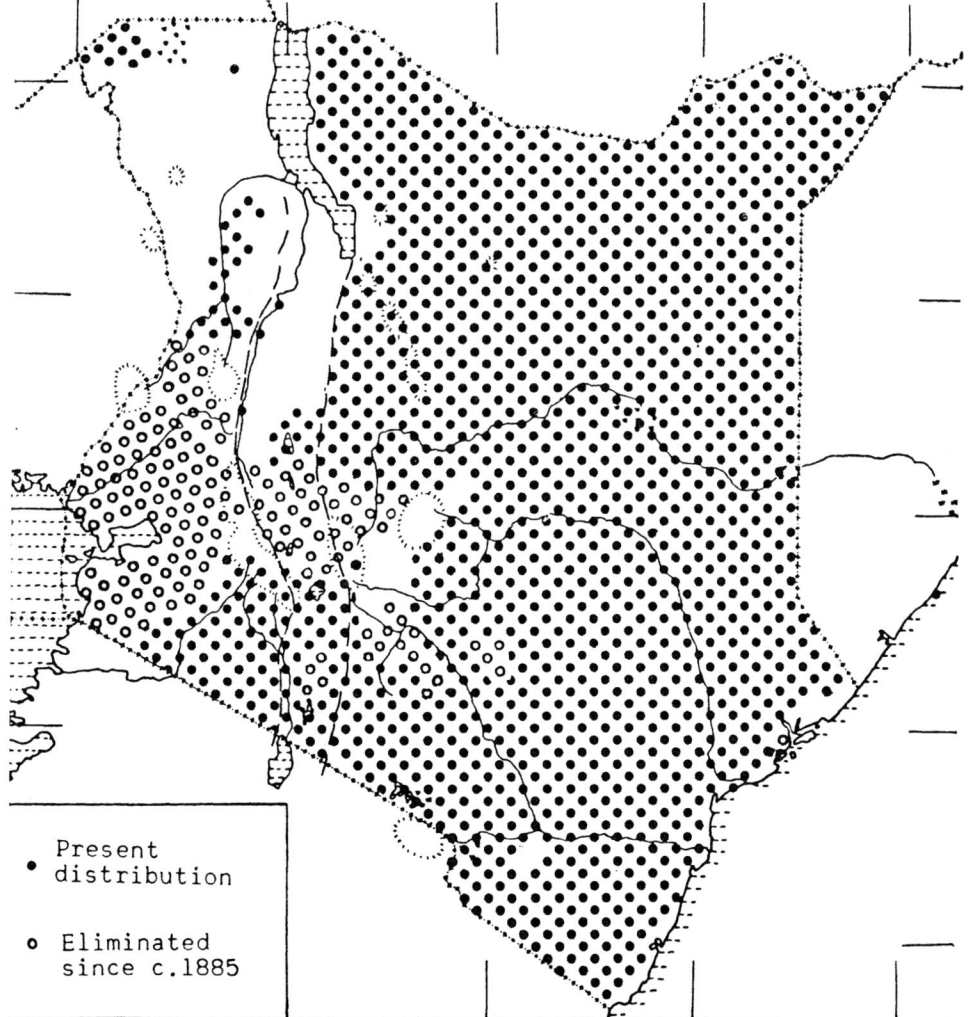

Fig. I–4
Lion (Panthera leo
Linnaeus)
(From D. R. M.
and J. Stewart
1963)

• Present distribution

o Eliminated since c.1885

from both lion and tiger. Guggisberg (op. cit.) believes that this large cave lion may have been succeeded and supplanted by the smaller, more adaptable present day lion whose picture appears simultaneously with that of the larger, maneless cave lion on Pleistocene cave paintings in Europe.

The present form of *Panthera leo* was distributed in historic times over practically the whole continent of Africa together with parts of Europe and Asia as far east as India. It became extinct in Europe at about the first century A.D. (Meyer, cited by Mazak, 1965) and has disappeared from North Africa and the Middle East during the last 100 years (Guggisberg, op. cit.). It is now found only south of the Sahara in Africa and in the Gir Forest wildlife sanctuary in India, where, in a 1280 km² area, a few hundred animals still survive.

Variously known in former times as the 'king of the jungle' and the 'king of the desert', the lion, in fact, inhabits many different types of surroundings, but is found neither in dense forest nor in complete desert. Its most frequent habitats are open grassy plains, woodland or bush, although it may penetrate into semidesert areas with scrubby vegetation.

The distribution map, Fig. I–3 shows that in the early nineteenth century it was spread all over the continent of Africa with the exception of the West African rain forests and the Sahara, where it occurred, up to the middle of the last century, only in the highland wilderness of Aïr.

While a fair number of lions still survive in Africa south of the Sahara, their range is gradually being restricted by the expansion of cultivation and their numbers are steadily decreasing. Fig. I–4 shows the changes in distribution in Kenya since 1885. The area where lions in the wild are tolerated is being gradually but inexorably reduced. In the opinion of many officials concerned with conservation, the time is near when the larger species of game animals will be restricted to the Game Parks. Thus it is all the more important to ensure that the management of these Parks is based on sound knowledge of the animals, their habits and their ecology.

11 Population Structure and Ranges

POPULATION SIZE, COMPOSITION AND DENSITY

The total number of lions in Nairobi Park remained substantially stable during the study period: there were twenty-eight animals in June 1968 and twenty-seven in June 1969. Table II–1 shows the sex and age composition at the beginning and at the end of the study year: the last column indicates the total number of all individuals encountered during this study.

Although emigration, immigration and births changed the individual composition of the population, age composition underwent no significant changes. Juveniles (up to approximately two years of age) comprised 57% of the population in 1968 and 62% in 1969, an insignificant difference. This is a much higher percentage of immatures than Makacha and Schaller report (1969) for the Lake Manyara National Park where the comparable figure is 20–25%

A large percentage of young is usually characteristic of a growing population. According to Wright (1960) a low average number of cubs per pride indicates that the population is nearing the limits of its food supply. The 2·7 immatures per pride in Serengeti in 1957 suggested to him that the population there was approaching such a limit. The average for the Nairobi

Table II–1
Sex and age class composition of population

Sex and age class	No.	June 1968 (per cent)	Sub-total	No.	June 1969 (per cent)	Sub-total	Total turnover No.	Per cent
Adult male	2	7·1	42·8%	1	3·7	37·0%	2	32·6
Adult female	10	35·7		9	33·3		12	
Juvenile male (12–24 months)	7	25·0	57·2%	2	7·4	62·9%	29	67·4
Juvenile female	1	3·6		3	11·1			
Cub male (up to 12 months)	4	14·3		7	25·9			
Cub female	4	14·3		5	18·5			
Total	28	100·0		27	99·9		43	100·0

lions is 4·4 immatures per pride, well above the one or two immatures cited by Wright for the Kruger-Sabi area at the time the lion population had outgrown its food supply (Stevenson-Hamilton, 1937, cited by Wright, 1960).

The preponderance of adult females over adult males throughout the study period is paralleled by the findings of Makacha and Schaller (op. cit.) at Lake Manyara Park; a similar situation was reported by A. Rogers in a lecture in 1967 at University College, Nairobi, on the game population of the Selous Game Reserve in Tanzania, and by Pienaar (1970) for the Kruger Park.

Both Pienaar (op. cit.) and Guggisberg (1961) suggest a higher mortality for male than female cubs, and according to the latter authority this occurs when the young males have to leave their pride at about two years of age, when they are not yet very proficient in providing for themselves. Another possible source of higher male cub mortality will be suggested later.

Population density remained stable during the period of this study (Table II–2); At 0·23 and 0·24 lion per km.^{2} it was comparable to the average density between 1961 and 1967 in this Park (Foster and Kearney, 1967) and to the figure given by Fosbrooke (1963) for Ngorongoro of 0·2 per km.^{2} In Lake Manyara Park the figure is almost double that of Ngorongoro and Nairobi, 0·38 (Makacha and Schaller, op. cit.). The authors suggest that the difference is related to the much higher biomass carried by that Park. Dowsett (1966) reports a lion density of only 0·12 per km.^{2} for the Ngoma area of the Kafue National Park in Zambia.

Table II–2
Density of lion in Nairobi National Park
Area: 114·8 km.^{2}

	1968	1969
Adult males	2	1
Adult females	10	9
Juveniles under 2 years	16	17
Total	28	27
Lion per km.^{2}	0·24	0·23
km.^{2} per lion	4·10	4·25

THE PRIDES: COMPOSITION AND CHANGES THROUGHOUT THE YEAR

Although mention is made in the literature of large prides numbering well over twenty animals (Guggisberg, 1961; Wright, 1960) and a pride of as many as thirty-five individuals has been reported from the Serengeti National Park (Schaller, 1969), the largest pride in this Park consisted of twelve animals: one male, three females and eight juveniles between one and two years old.

It is possible that the larger prides encountered in some areas are an adaptation to the presence of scavengers, especially hyaenas, occurring in large packs. Smaller prides take longer to consume a given prey animal and thus are more likely to be molested by hyaenas. In competition for a carcase, superiority depends on the relative number of lions and hyaenas (Kruuk, 1966; Schaller, 1969; Cullen, 1969). Cullen quotes several accounts from Tanzania Parks archives when hyaenas in large packs succeeded in driving lions off their kill. In one case eleven lions, eight of which were immature, were driven off by a pack of eleven hyaenas, in another instance 'two lions on a zebra kill were vanquished by a mob of twenty-nine hyaenas which managed to get the

Table II–3
List of prides and
individuals in each
pride

Pride	Individuals	Sex	Approx. age
1. Athi I	Kihara	m	adult
	Sophie	f	old
	Nike	f	adult
	Fathead	f	adult
	Sally	f	juvenile*
	YM 1	m	juvenile*
	YM 2	m	juvenile*
	YM 3	m	juvenile*
	C 7	m	juvenile*
	C 9	m	juvenile*
	C 10	m	juvenile*
	C 11	m	juvenile*
2. Athi II	Nike	f	adult
	Bertha	f	adult
	Sally	f	young adult
	Sosian	m	juvenile*
	Kyra	f	juvenile*
	Tamara	f	juvenile*
3. Romola	Romola	f	adult
	Patricia	f	adult
	Chryse	f	young adult
	Calef	m	juvenile*
	Karen	f	juvenile*
	Carla	f	juvenile*
	Victor	m	juvenile*
	Pamela	f	juvenile*
4. Misty	Misty	f	adult
	Anne	f	young adult
	Elgon	m	juvenile*
	Helen	f	juvenile*
	Tina	f	juvenile*
5. Lassie	Lassie	f	adult
	David	m	juvenile*
	Batian	m	juvenile*
	Nelion	m	juvenile*
6. Scarface	Scarface	m	adult

kill for themselves.' Thus bigger prides could have definite survival value in areas where hyaenas occur in large packs, just as hyaenas are believed to exert selective pressure in favour of larger packs in wild dog populations (Estes and Goddard, 1967).

In Nairobi Park the hyaena population has been negligible for the last seven years at least. Foster and Kearney (1967) reported that none were seen during the 1966 game counts, although two families were believed to be resident. During my study, hyaenas were encountered only three or four times and no more than two individuals were ever seen together. However, K. de P. Beaton (1949) speaks of a pride of sixteen in the Park in 1949 and at the same time relates an incident in which seven or eight hyaenas followed and molested a lion carrying part of a carcase and succeeded in appropriating the meat from the cat. This may possibly suggest that when hyaenas were more numerous, prides were also larger.

In June, 1968, there were four prides in the Park, and a solitary male known as Scarface (Table II–3). This male appeared in the Park towards the end of 1967 and seems to have displaced a resident male known as Spiv

(Guggisberg, pers. com.) who disappeared shortly thereafter. Whether any members of Spiv's pride remained behind is not known. This male, Scarface, as will be shown in more detail below, was subsequently associated with all the resident prides in turn, but could not be considered as part of any one of them. He will occasionally be referred to as 'the male' as from July, 1968, onwards he was the only resident male in the Park.

The largest pride in June, 1968, Athi I, consisted of twelve animals (Table II–3 and Fig. II–1). The adult male and one adult female disappeared in July 1968 and had not been seen again by the end of this study. Another female, presumed to be the mother of some of the juveniles, together with the seven male juveniles, was last seen in February 1969. As the longest interval between sightings before then was two months, they were presumed to have left the Park, at least temporarily. Two females of this pride (one classified as juvenile at the beginning of this study reached sexual maturity during the period and was seen mating in May 1969) linked up with a solitary adult female (first seen 30th September 1968) to form a pride designated as Athi II.

The Misty pride, consisting of three females and four cubs at the beginning of this study, broke up after one other female, Lassie, produced a litter of three cubs and did not rejoin her pride after the customary isolation period before and after parturition. This period usually lasts until the cubs are four to eight weeks old (Schenkel, 1966; Kühme, 1966).

The Romola pride was the most stable unit throughout this year. It consisted of three females and two litters of cubs. Towards the end of the study period the third female also produced cubs which, by the end of the study, were not yet integrated into the pride. The mother, however, was still considered part of the pride as she was seen associated with it on most occasions when it was observed.

Fig II–1
Composition of
prides throughout
study period

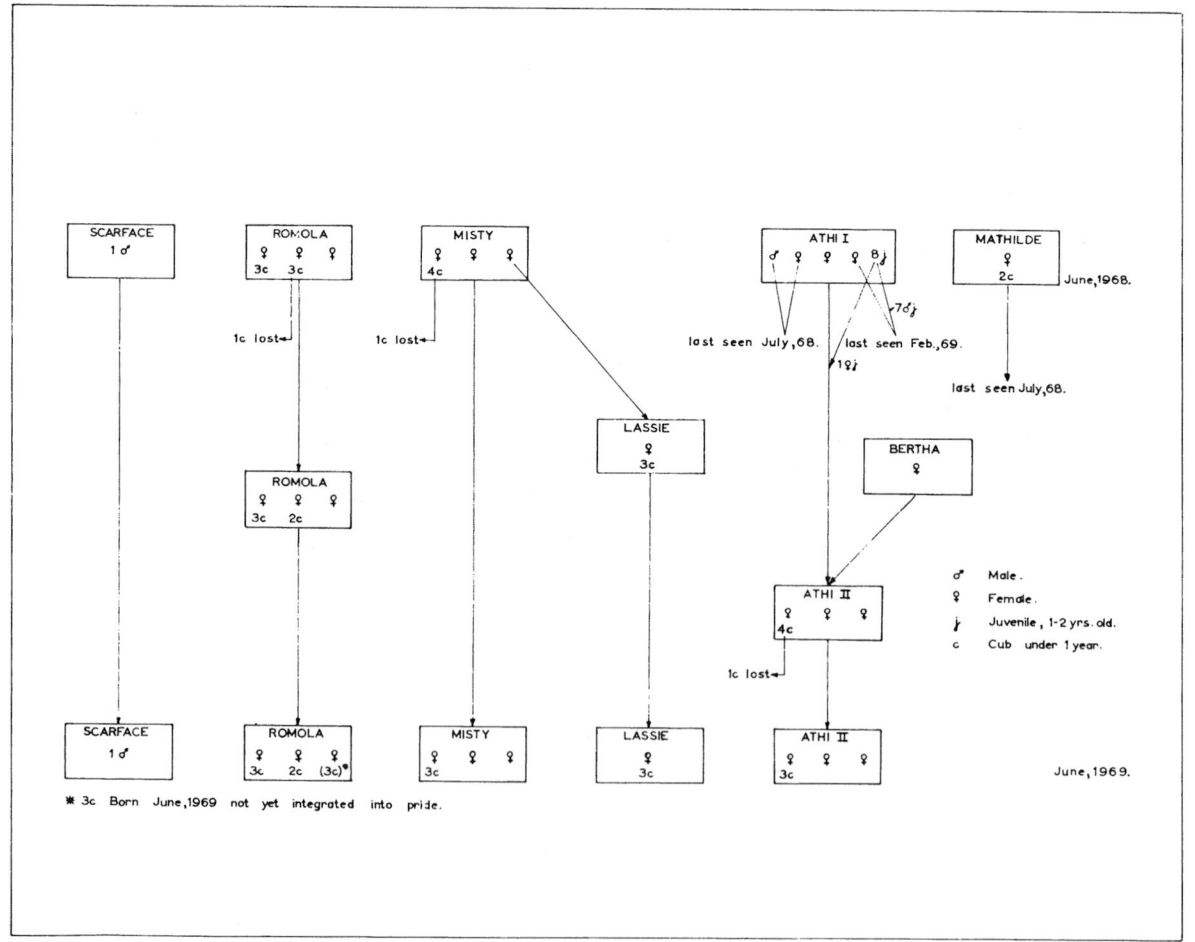

The fourth pride present in June 1968, a female with two cubs, was not seen after July 1968 and was presumed to have left the Park.

Thus, in June 1969, there were again four prides and one solitary male, although, as seen from the foregoing account, considerable changes had occurred.

HOME RANGES

Most animals spend the majority of their time within a definite area usually called a 'home range'. This designation does not necessarily imply exclusivity as the term 'territory' does when an area is kept free of all, or a certain class of, conspecifics. An animal's territory may include all of its home range, or, more frequently, constitute only a certain, defended, part of it.

For the purpose of this discussion I shall call the area within which any member of a pride was encountered during the study, the home range of that pride. Home ranges will be considered under two aspects, day and night. This is partly because night observations were sporadic and intermittent and could thus not be compared with day observations on an equal basis, and partly due to the fact, discussed below, that utilization of areas showed a different pattern by day and by night.

In calculating the day range of a pride all positions where any member of the pride was seen between 0700 hours and 1900 hours any one day were marked on a map with a 1 km.² grid. The outermost points were then connected by straight lines to give the day range for that pride. This is the 'minimum area' method of calculating ranges which assumes that the outermost points of occurrence define the limits of the area 'utilized by the animal during its normal activities 'using Burt's (1943) definition of home range (Van Vleck, 1969).

As shown on Fig. II–2, the yearly ranges overlap to a large extent. This is a similar situation to that found in wolf and hyaena ranges (Schaller & Lowther,

Fig. II–2
Day ranges—all
periods combined

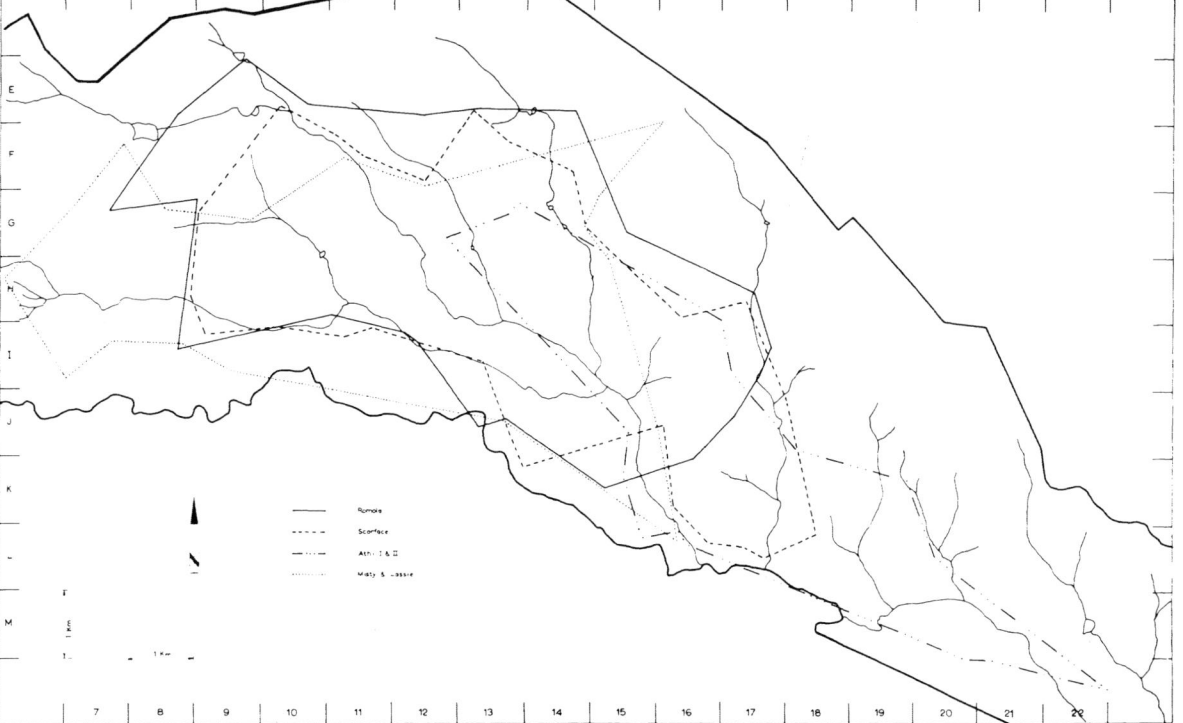

1969). (For this figure the ranges of Athi I and II are combined as are the ranges of Misty and Lassie.) In broad outline it is possible to say that the Athi pride occupied the most easterly portion of the Park, Misty the westernmost section and Romola and Scarface the more central area. The range of every pride overlapped to some extent the range of every other when taken on a yearly basis. However, when more limited time intervals were considered there was a tendency toward an increasing separation of the prides during the year.

Figs. II–3, 4, 5 and 6 show the ranges broken down into three periods:
  (i) Sporadic observations between 1st January and 30th May 1968 and regular observations between 1st June and 30th September 1968;
 (ii) Regular observations between 1st October 1968 and 31st January 1969;
(iii) Regular observations between 1st February and 15th June 1969.

The spatial and temporal relationship of the five ranges with one another will be examined in turn. Athi I and II will be considered together but, for reasons that will be clear later Lassie and Misty are taken separately for Period III.

Scarface and Romola: These two ranges showed the greatest degree of overlap seen in all the possible combinations. However, there was a closer concordance in range in the first two periods than in the third, when the range of the male extended further to the southwest and southeast. Both ranges, showed a considerable increase in size between Period I and II and little increase between Period II and III. That of Romola had a more compact shape, while Scarface's showed two extensions in the third period from a central area: to the southwest and southeast. These were the two areas where his range coincided with those of the Athi and Misty prides.

Fig. II–3
Day range—
Scarface

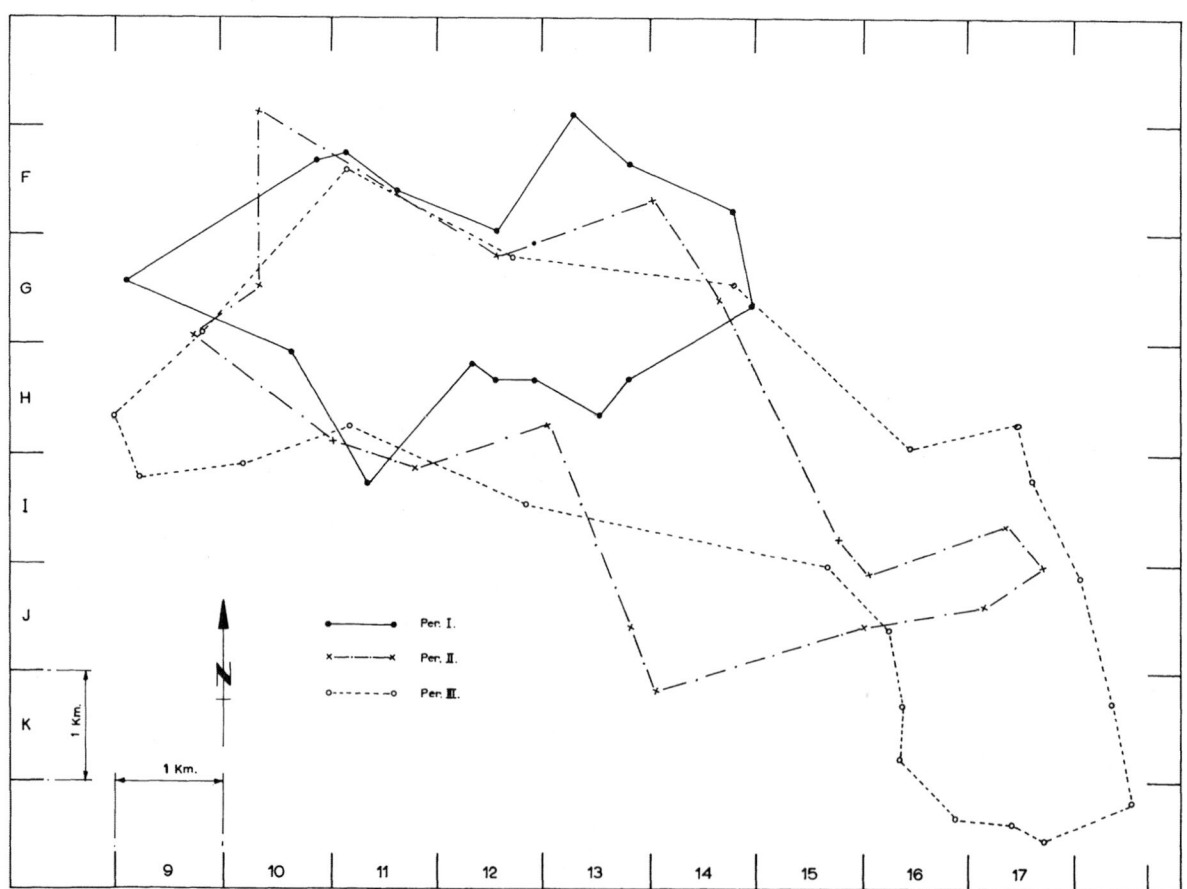

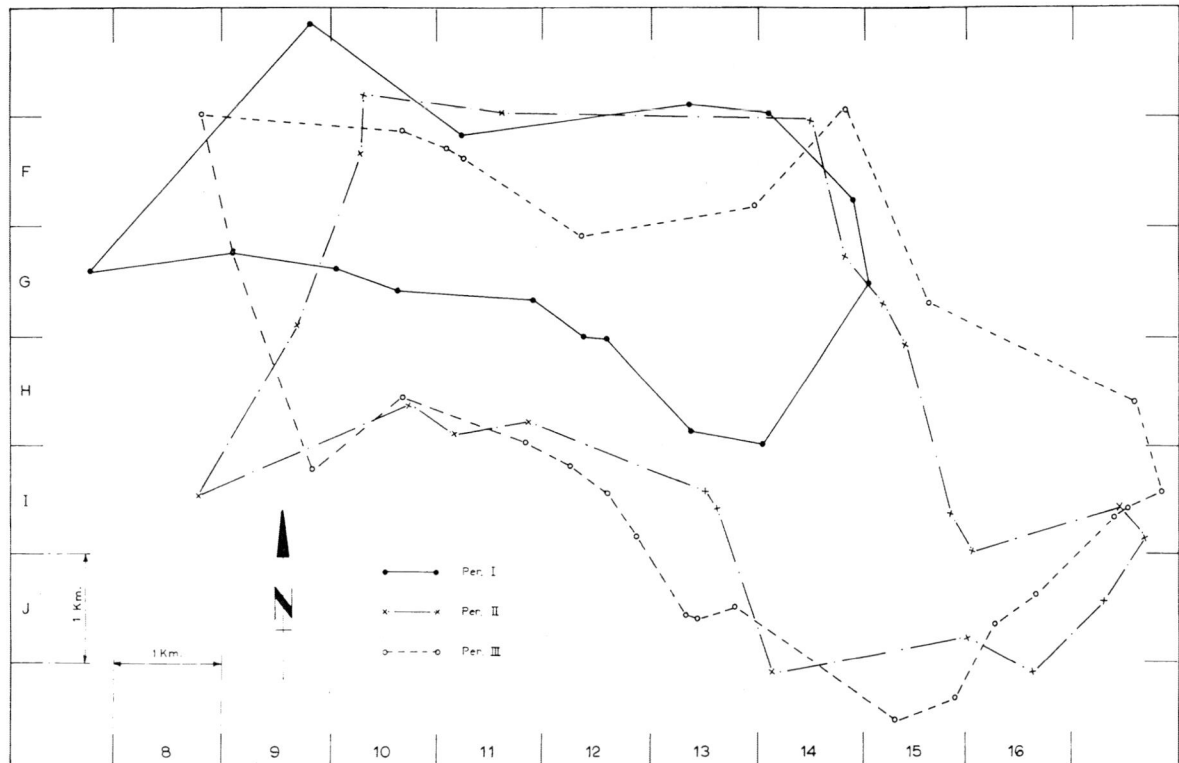

Fig. II-4
Day range—
Romola

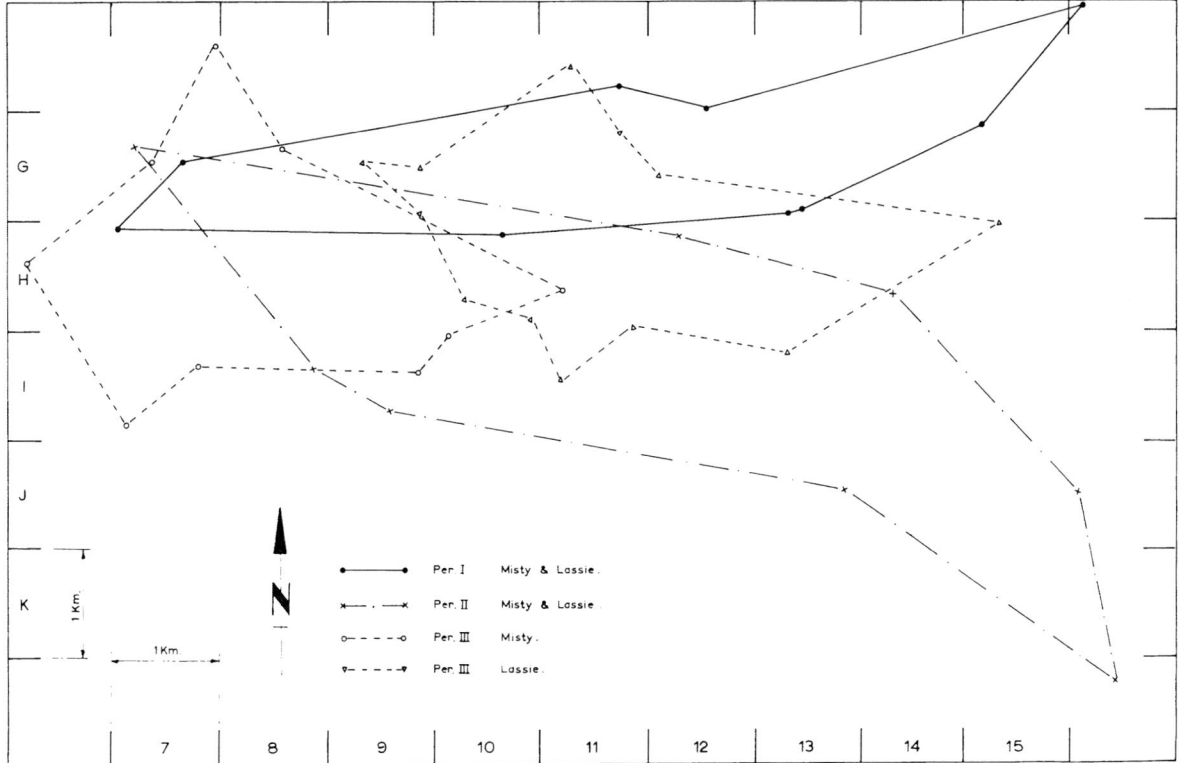

Fig. II-5
Day range—Misty
and Lassie

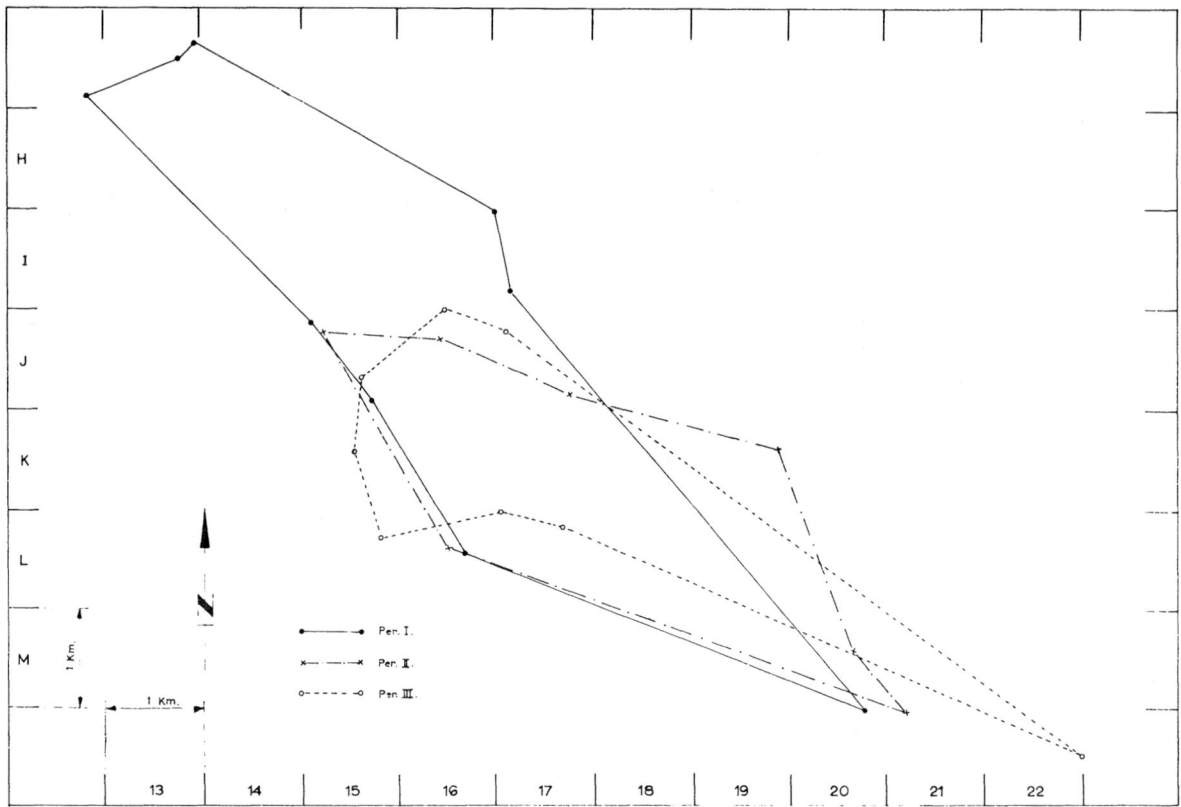

Fig. II–6
Day range—Athi I
and II

Scarface and Athi: As seen in Table II–3 at the beginning of this study the Athi pride had an adult male, Kihara, who was resident in the Park before Scarface appeared. He was last seen in July, 1968. This may account for the fact that during the first period there was only a very slight overlap between the ranges of Scarface and Athi. In the second period Scarface extended his range towards the southeast by 3–4 km. but the Athi range contracted at the same time and to the same extent so that the ranges continued just to overlap. In the third period there was little change on the western side of the Athi range, while it extended somewhat further to the southeast; but at the same time Scarface's range extended another 3·5 km. further southeast thus covering about half of the area of the Athi range and including all the core areas.

Scarface, Misty and Lassie: In the third period when Lassie and Misty had separate ranges, the male's range covered most of Lassie's but Misty's extended well beyond Scarface's to the southwest; less than one-third of her range overlapped the male's.

Romola and Lassie: Lassie's range (Per. III) was almost entirely contained within Romola's. As will be shown in Chapter V, there was an occasional, but never a close, association between these two prides. In Period III Lassie's range was more compact than in the first two, possibly due to the fact that she had young cubs (less than seven months old) during this time. This may be the reason why she established a range of her own. Romola's range, on the other hand, showed the opposite trend, expanding during the second and third periods. Again, as in Lassie's case, the period of the smaller range coincided with the time the cubs were less than six months old.

Romola and Misty: For the first two periods the two ranges overlapped to a great extent; in the third, Misty's was almost entirely beyond Romola's towards the west-southwest.

Romola and Athi: In the first two periods the situation here was similar to that between Scarface and Athi. In the third period Romola retreated somewhat from her eastern boundary. Athi retreated a little from its western border, while extending more towards the southeast. The changes on both sides were enough to effect a separation.

Lassie and Athi: There was a slight overlap during the first two periods; during the third, boundaries were more than 3 km. apart.

Misty and Athi: For the first two periods as above; in the third boundaries were 6 km. apart.

Misty and Lassie: For the first two periods Lassie was part of Misty's pride; the separation occurred by the end of January 1969. From 1st February to 15th June (Per. III) the two ranges overlapped very little, with Misty's extending more towards the southwest and Lassie's occupying a more central location.

All these results indicate that during the year a consolidation of positions took place in the spatial distribution of the prides. If we consider all the prides' relationships to each other with the exception of their relationship to the male, we find the same picture in every instance: in the third period the separation was always greater than in the first two.

The male's original attachment was to Romola's pride and that pride could be considered his primary association: throughout the whole study period his relationship with that group was closer than with any of the others (see Table II–4). Early in the study the Athi pride had its own resident male, which presumably prevented Scarface from extending his range towards the southeast. However, after the disappearance of the Athi male, Scarface was able greatly to extend his sphere of activities. During the last four months he was seen associated with the females of the Athi pride on several occasions.

The least degree of association is indicated by Scarface's relationship with Misty, as his range in the third period covered less than one-third of Misty's. Table II–4, based on the number of observed associations, rather than on coincidence of ranges, confirms this: in only 2% of all observations was he associated with Misty's pride.

This situation, where the male is not an integral part of any one pride, but associates in turn with various prides of females and cubs, is similar to that observed in Lake Manyara Park by Makacha & Schaller (1969) where two males ranged over an area occupied by two distinct prides of females and cubs and divided their time about equally between them. A similar pattern of range occupancy and association was reported by Guggisberg (1961) for the Nairobi Park, where the range of two associated males included two prides of females.

Table II–4
Scarface's
association with
various prides

Prides Scarface was seen with	I		II		III		I, II, III combined	
	Number of obs.	*Per cent*	*Number of obs.*	*Per cent*	*Number of obs.*	*Per cent*	*Number of obs.*	*Per cent*
Romola	36	100·0	27	93·1	13	40·6	76	78·3
Athi II	—	—	1	3·5	12	37·5	13	13·4
Misty	—	—	—	—	2	6·3	2	2·1
Lassie	—	—	1	3·5	5	15·6	6	6·2
Totals	36	100·0	29	100·0	32	100·0	97	100·0
Scarface seen alone	8		6		10		24	

I Period from February 1968 to 30th September 1968.
II Period from 1st October 1968 to 31st January 1969.
III Period from 1st February 1969 to 15th June 1969.

Schaller (1967) reported a situation in Kanha, India, where one adult tiger apparently shared his range with three resident tigresses. Schaller believes that the male in this case defended his range only against other males but shared it willingly with females.

Day range size did not increase in direct proportion to numbers in a pride. Scarface alone ranged over an area of 29 km², only slightly less than the 29·9 km² of Misty, with five animals and the 31·2 km² of Romola with eight animals. However, as only day resting positions were considered in this calculation of range, it may not be an accurate measure of the area required for the maintenance of a pride. As their hunting activities often took the prides outside their day range (see Chapter III), a combined hunting-and-day range would give a truer picture of the actual maintenance range of the prides. But as night observations were too scanty, no meaningful calculation of night ranges could be made. The available observations indicated, however, that a much larger area was actually utilized than observations made by day only would suggest.

Various range sizes for lion found in the literature are: 52·8 km² (Wright, 1960), 20·8 km², 12·8 km², 86·4 km² (Guggisberg, op. cit.) for Nairobi Park and 63·23 km² and 55·30 km² for Kruger Park (Pienaar, 1970). Figures for this study range from 19·75 km² to 31·19 km² (Table II–5, Column V).

For comparison, ranges of tigers are reported by Schaller (op. cit.) in Kanha as 40 km² and 48 km², while figures quoted by the same author from the literature range from 14·4 km² to as much as 2400 km², the latter from Corbett (1957) relating to a man-eating tiger. Thus range sizes for tigers seem to be much more variable than for lions, although as yet information on both species is sparse.

DAY RANGE UTILIZATION

As we have seen, all ranges overlapped to some extent, if taken on a yearly basis. However, during the study a gradual spacing-out occurred with the ranges tending to become more compact and overlap being reduced. This spacing-out was also evident in the pattern of utilization of the range by the prides.

Ranges are not used uniformly and certain centres of activity or centres of maximum usage can be established (Linn, 1965). These may be areas of importance because they offer favourable resting or hiding places, or because they are the best areas for feeding (Ewer, 1968). The designation of 'core areas' used by Ewer is adopted here for these centres of maximum utilization. In this case maximum usage implies the use of the areas as resting places by daytime.

To establish these core areas a method adapted from Bromley (1969) was used. All daytime positions were marked on a separate map for each pride, then a 1 km² square was drawn to scale on transparent paper. This square was moved over the map: when it covered an area containing more than 10% of the daytime positions, this was designated a primary core area; when it covered an area containing between 8% and 10% of positions, it was designated a secondary core area.

In Table II–5 the number and distribution of the daily positions and the core areas is analyzed, to show the degree of concentration of usage. All core areas were added (Column I) and expressed as a percentage of the total day range area of that pride (Column VI). All positions within the core areas were also added (Column II) and expressed as a percentage of all positions for the pride (Column IV) (Ables, 1969). These two percentages were used to express the degree to which usage was concentrated, inasmuch as the largest percentage of positions within the smallest percentage of area re-

Pride	I No. of km.² with over 8⁰₀ of total positions	II No. of pos. in I	III Total pos. for pride	IV II as percentage of III	V Total area of day range km.²	VI Area in I as percentage of V	VII IV/VI	VIII Ratio of concentration
Scarface	3	25, 18, 11 } 54	128	19·5, 14·1, 8·6 } 42·2	29·0	10·3	4·1	over 4
Romola	5	37, 31, 23, 22, 18 } 131	221	16·7, 14·0, 10·4, 9·9, 8·2 } 59·2	31·2	16·0	3·7	between 3·5 and 4
Misty	3	8, 5, 5 } 18	42	19·0, 11·9, 11·9 } 42·8	29·9	10·0	4·3	over 4
Lassie	3	24, 6, 6 } 36	64	37·5, 9·4, 9·4 } 56·3	22·3	13·4	4·2	over 4
Athi I and II	4	13, 6, 6, 5 } 30	46	28·3, 13·0, 13·0, 10·9 } 65·2	19·7	20·2	3·2	between 3 and 3·5

Table II–5
Day range
utilization.
Concentration of
usage in relation to
range size

presents the highest degree of concentration (Column VII and VIII).

By these criteria the strongest concentration was found in the ranges of Scarface, Misty and Lassie. In the latter case it was primarily due to the fact that during the later months utilization of her day range was very heavily concentrated within 1 km.² resulting in the highest degree of utilization for any km.² for any pride. 37·5% of all recorded daytime positions for this animal fell within 1 km.²

It is usual to find that, while ranges overlap, core areas do not, or only to a very slight degree (Ewer, 1968; Schaller and Lowther, 1969). The Nairobi Park lions show just such a pattern. Fig. II–7 represents the core areas of all prides. A similar spacing out is noticeable here, as is evident from the periodic day range figures. Denser shading denotes primary, less dense, secondary core areas.

Primary centres were generally well spaced out, except in the case of Scarface and Romola. As all other data show, the male was most closely associated with this pride; the almost complete overlap of two primary and one secondary core areas confirms this. There was only one other case of a very slight overlap of primary core areas (12·5% overlap) in the case of Lassie and Romola and Scarface. However, Lassie had a certain, even though tenuous, relationship with the Romola pride, especially towards the latter part of the year.

Secondary areas show very much the same picture with overlap only in the case of Scarface, Romola and Lassie.

HABITAT PREFERENCES

The positions of the daytime resting places for all the prides were primarily clustered along the watercourses (Fig. II–8). Out of a total of 501 positions recorded, only twenty-six (5·2%) were further than 500 m. away from a

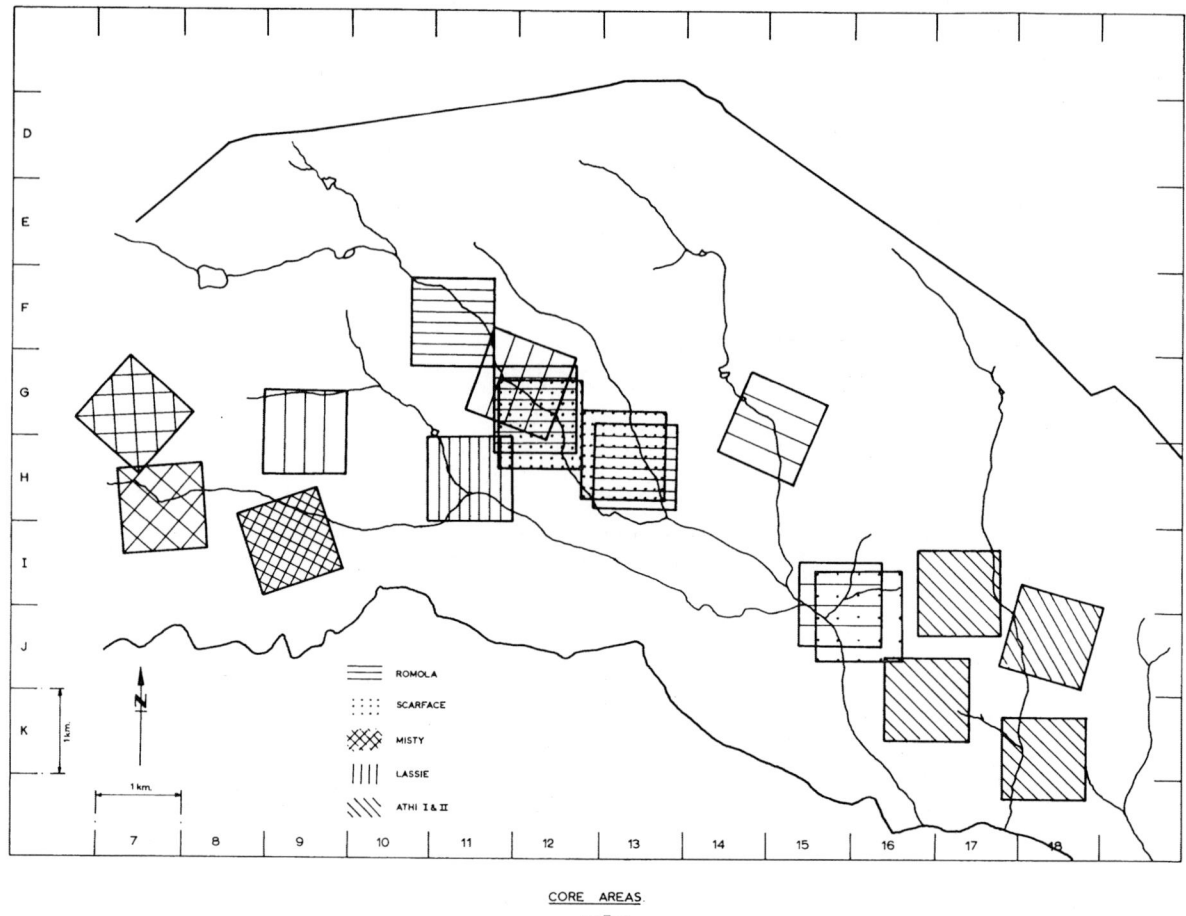

CORE AREAS.

FIG. II – 7

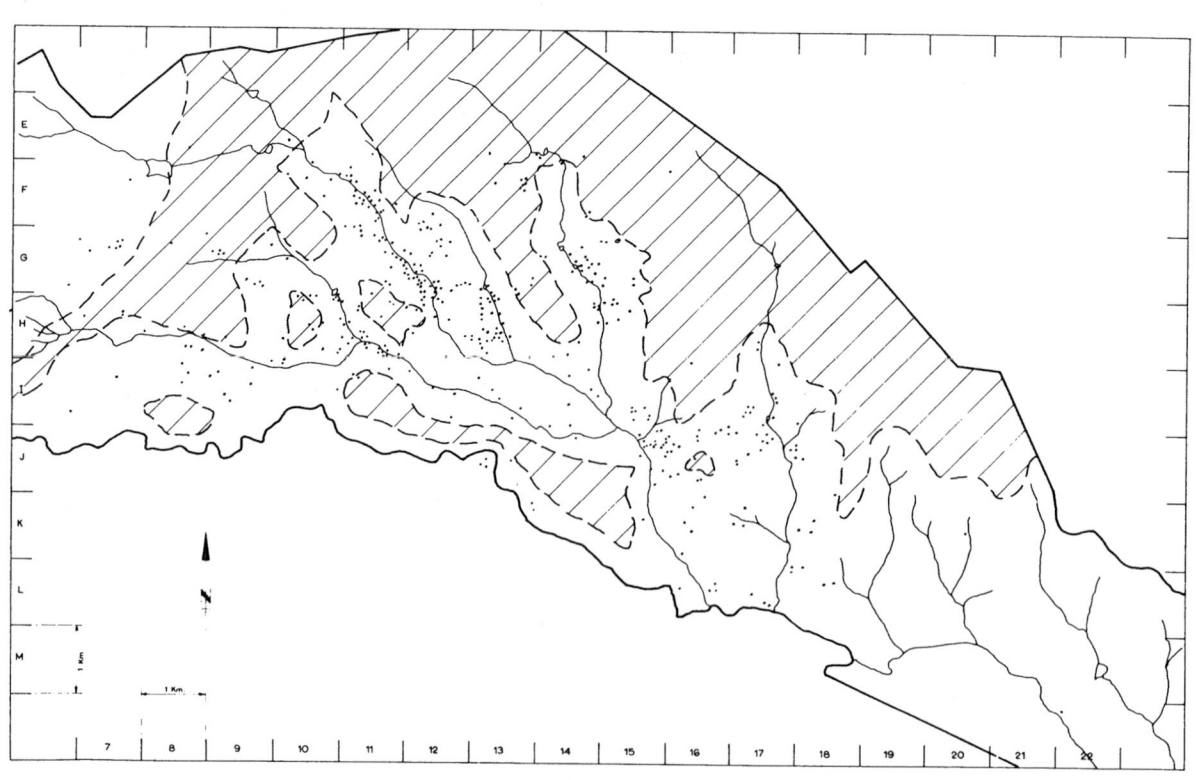

DAYTIME POSITIONS — ALL PRIDES.

FIG II – 8.

watercourse or dam, the former being much preferred.

Although the Park as a whole can be considered an area well endowed with water, there are, nevertheless, large tracts where watercourses or dams are very widely spaced. A look at Fig. II–8 will reveal that most of these areas are *Acacia drepanolobium* scrub. (Areas with diagonal shading; after Heriz-Smith, 1962). This study indicates that these areas were not normally selected by the lions for daytime resting positions, but were more extensively utilized by night for hunting. In most cases when animals were resting by day in *Acacia* scrub, they were resting near a carcase, on which they were intermittently feeding throughout the day or even over a period of several days.

The typical positions selected by the lions for their daytime rest periods had the following features in common:

(1) less than 500 m. distance from water, preferably a river course;
(2) proximity of scattered bush or riverine thicket for cover;
(3) some elevated point such as termitarium, rock outcrop or murram heap;
(4) substrate: short grass or rock.

The *Acacia drepanolobium* areas, as a rule, do not provide any of these conditions. Large tracts have no watercourses, and although the *Acacia drepanolobium* trees do provide shade, they do not provide cover. There are very few termitaria in these sections of the Park and scarcely any rocky outcrops, few short grass areas and no rocky substrate.

Another possible reason for the animals' avoidance of these areas by day may be that these sections of the Park were extensively used by night for hunting. In fact, 36% of all recorded kills were made in *Acacia* scrub. So it may be to the predators' advantage not to disturb the prey animals in these sections by their presence during the daytime.

Contributing factors to this pattern of space utilization may be the following. Lions require water after feeding, although they were not observed to drink frequently at other times. (See Chapter VII). When the pride finished feeding, it usually left the kill immediately, regardless of the time of day or night. In most cases it then travelled to the nearest body of water to drink. After the animals had quenched their thirst, they settled down in the vicinity. But, if the neighbourhood of the nearest source of water did not provide most of the features listed above, they tended to move on to settle down in a more suitable locality.

III Activity patterns

DAILY ACTIVITY CYCLES

Lions are usually nocturnal, concentrating their activities largely within the hours between sunset and sunrise (Kühme, 1966; Estes, 1967; Kruuk and Turner, 1967; Schaller, 1969). It has been suggested that where lions are not hunted, they may turn more towards diurnal habits but this does not seem to be the case in the Nairobi Park population. It may be that disturbance by visitors has the same effect as disturbance by hunting, but this is merely a speculation.

The Nairobi lion population can definitely be considered nocturnal in that most of their active period falls between 1900 hours and 0700 hours, roughly between sunset and about one hour after sunrise. Their activity cycle will be examined under two different categories, daytime activities and night activities.

The figures for the daytime are more detailed and accurate, as there were only a limited number of nights each month when observation was allowed by the Park authorities and even then the lions could not always be followed throughout the night. Also, identification by night was more difficult and, while the identity of the pride was known in every case, individual animals were not easy to tell apart.

DAYTIME ACTIVITIES

The activity pattern of the lions was analysed by dividing the population into three groups, Male, Female and Cubs (all animals under two years of age, approximately). Fig. III–1 shows three graphs for the three groups indicating the amount of activity as against rest, between 0700 hours and 1900 hours. Activity was recorded by minutes; for figures showing hours of observation, the totals were corrected to the nearest half hour.

All three graphs show the same general trend, a steep decline in active periods between 0700 hours and 0900 hours, when prides usually settle down after the night's hunting activities. There is a corresponding rise in activity beginning between 1700 hours and 1800 hours.

For further analysis all activities were grouped into six categories:
(1) Feeding, including suckling and drinking. Stationary.

(2) Hunting: here defined as necessitating fixed gaze in direction of prey animal, tense body and movement in the direction of gaze. Requires at least some displacement.

(3) Grooming: self or social grooming. Stationary.

(4) Walking: moving in a definite direction, when not hunting.

(5) Playing: solitary or social. Stationary or with displacement.

(6) Other: elimination, standing, greeting, etc. Every action not classifiable under (1) to (5).

(1) and (3) are stationary, (2) and (4) require displacement, (5) and (6) may be either.

Figs. III–2, 3 and 4 show these six groups of activity, each expressed as a percentage of total activity for that hour for the particular group of animals.

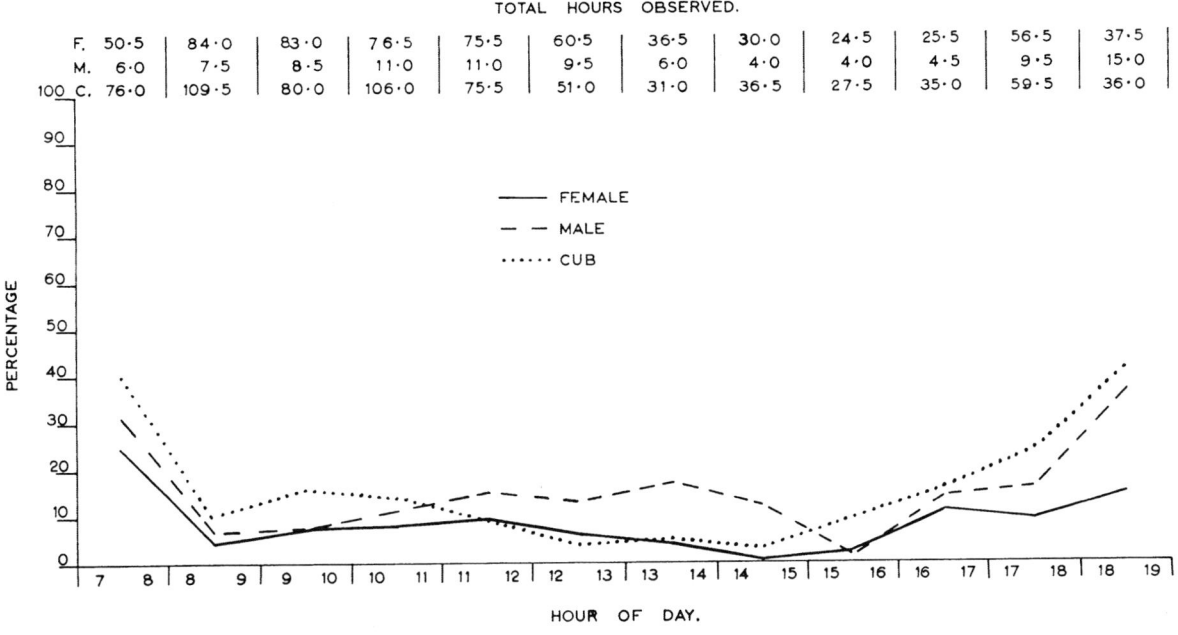

Fig. III–1
Daytime activity—
females, male and
cubs

ACTIVITY PATTERN OF FEMALES

Peak resting periods occur for the lionesses between 1400 hours and 1500 hours (Fig. III–2). The same hour also shows the lowest percentage of total activity: less than 10% (Fig. III–1). Of the small amount of activity that does occur, grooming takes up 75% and walking and 'other' 12·5% each.

The graph for hunting does not show any consistent pattern except for a low point coincident with the hour of least activity. Hunting, as here defined, may occur at any time during the rest period, but is less likely when the greatest number of individuals are in a state of maximum relaxation, that is, when combined alertness of the pride is at its lowest. This type of hunting activity differs from that dealt with in the section on night activities, in that it is always triggered off by a prey animal or animals coming into the visual field of the resting lions.

Two small peaks occur for play. They coincide with two of the four highest peaks for play on the cubs' graph. This is to be expected, as females rarely play alone or with other adults, but mostly with cubs. Cooper (1942) found that captive females continue to indulge in play up to about ten to twelve years old, while males rarely, if ever, play after they reach the age of four.

The highest peak for play on the cubs' graph occurs between 1800 hours and 1900 hours, when the females already begin to move about in preparation for the night hunting, as is shown in the sharp upswing in walking between 1700 hours and 1800 hours. From 1700 hours on, females start emerging into the open, if they have been resting in cover, and begin to move away by stages, walking away from the resting place for a few metres, then resting again for a short time, moving on again, and so on, until between 1800 hours and 1900 hours as a general rule, they start definitely to walk away.

Fig. III–2
Activities as a percentage of total activity per hour— females

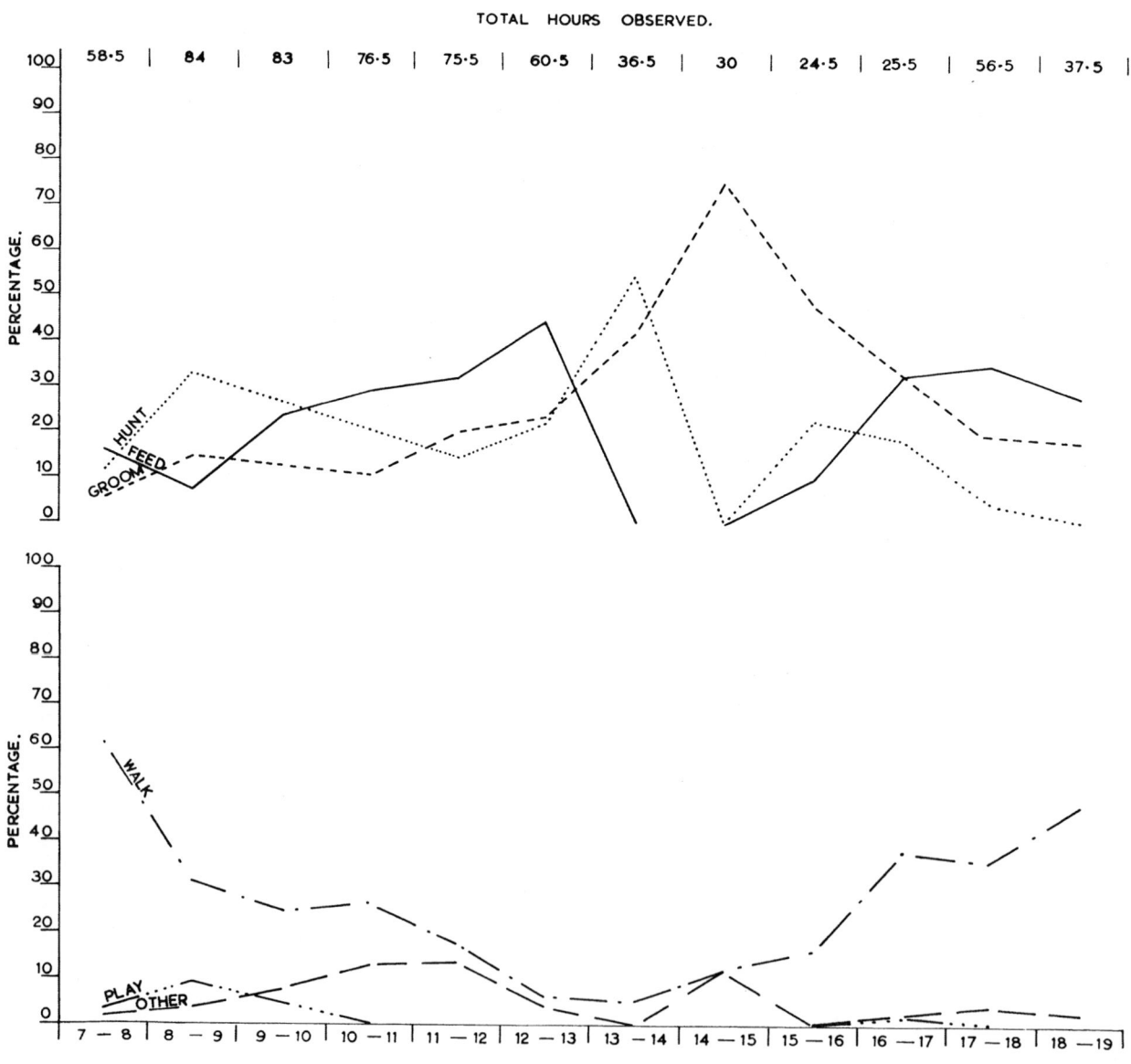

TOTAL HOURS OBSERVED.

ACTIVITY PATTERN OF MALE

The activity graph for the male (Fig. III–3) shows a greater degree of fluctuation than that for the females.

There is a striking difference between his graph for walking and that of the females'. Whilst the females' walking curve decreases sharply between 0700 hours and 0900 hours, that of the male rises continuously to a high between 0900 hours and 1000 hours before falling off sharply. This may be explained by the following. It was often observed that the male accompanied one of the prides during the night, usually keeping to the rear and following the females' movements with a short time lag. In the morning they usually separated, the females going to rest within their day range and the male within one of the preferred sections of his day range. These two centres of preferred occupancy only coincided in the case of Scarface and Romola, so in all other cases this necessitated a certain amount of walking for the male to get from the resting

Fig. III–3
Activities as a
percentage of total
activity per hour—
Male

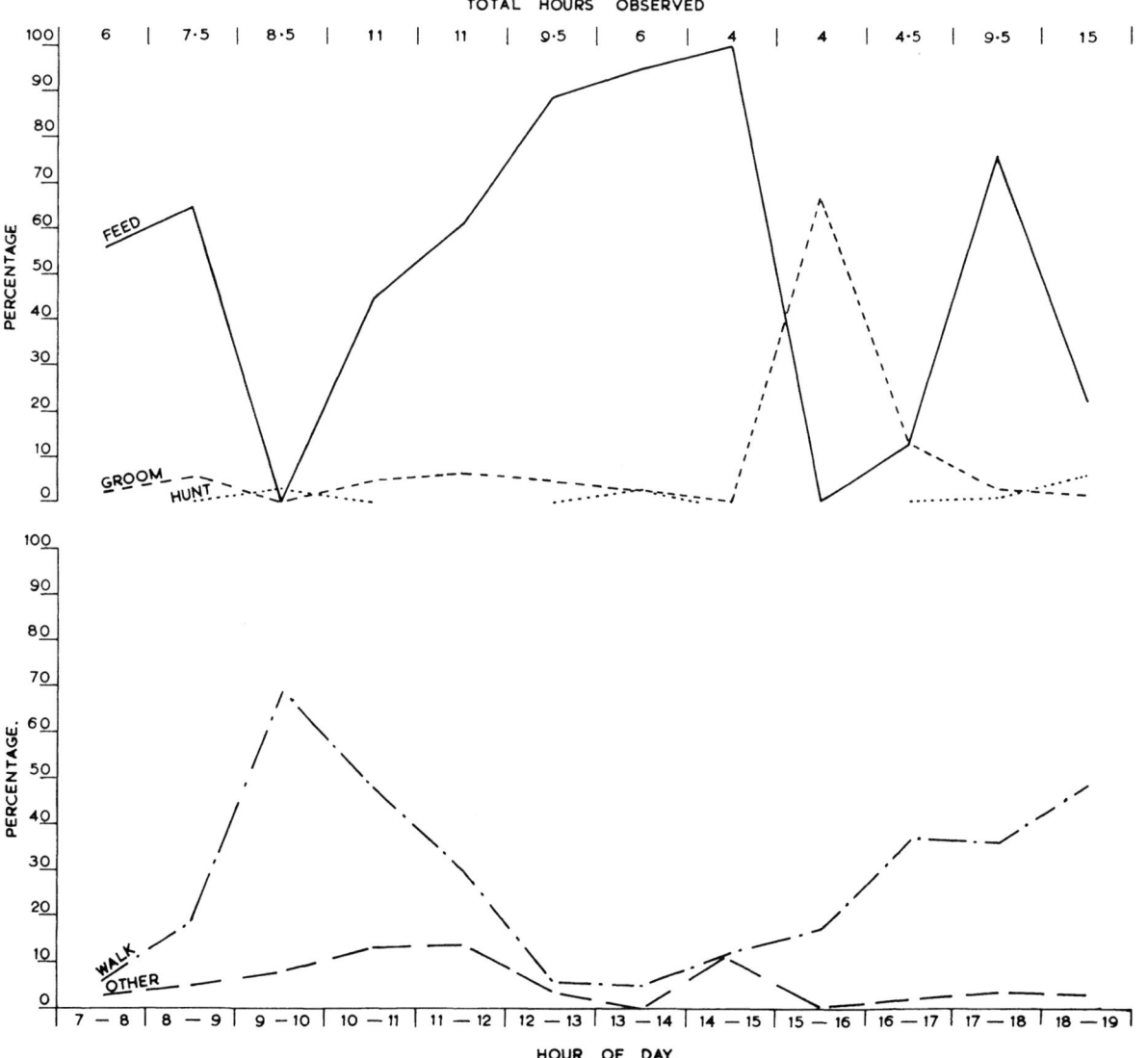

place of the pride he was accompanying by night to one of his own preferred places.

The male's curve also differ markedly from the females' in the percentage of hunting activity. Hunting never constituted more then 6% of any hour's activity for the male while it rose to 55% for the females. This reflects the fact that this male seemed to do very little hunting of his own, but was seen sharing in the kills of all the prides. Similar observations on males were made by Estes (1967) in Ngorongoro and by Schaller (1969) in the Serengeti Park.

The total absence of play from the male's activities is in keeping with Cooper's (1942) observation on adult males in captivity.

CUBS' ACTIVITY PATTERN

Fig. III–4
Activities as a
percentage of total
activity per hour—
cubs

The graph for active hours follows the general trend of the females' graph but at a higher level (Fig. III–4). This is not surprising, as younger animals have

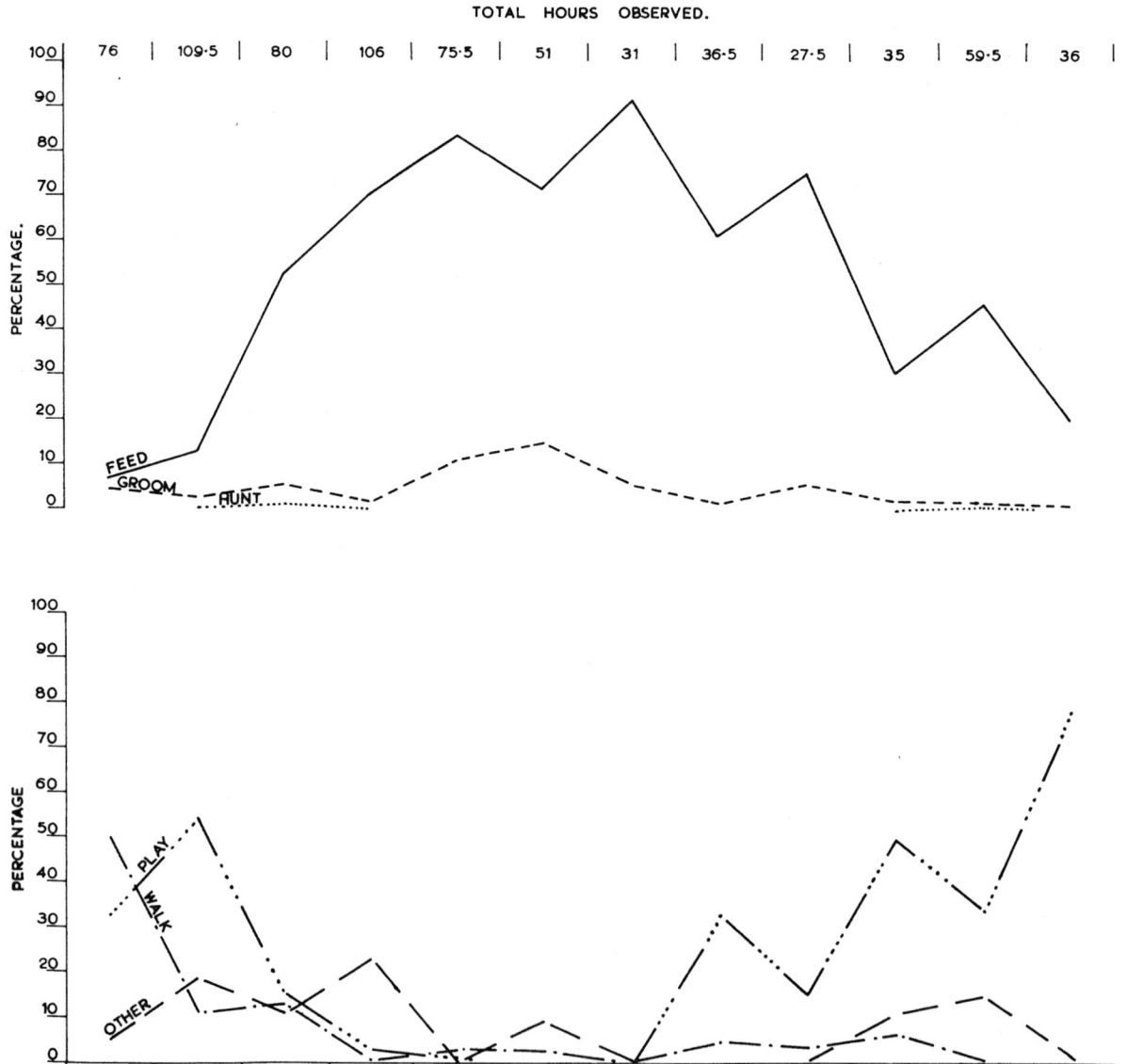

TOTAL HOURS OBSERVED.

HOUR OF DAY.

a higher level of activity than adults (Zuckerman, 1932).

As is to be expected, hunting constitutes only a very minor proportion of the cubs' activities and grooming is also on a very low level compared with the females. The gradual development of these two activity patterns in cubs will be discussed later, in the Chapter on 'Reproduction and Development'.

The two major activities are feeding and playing. These two activities show a largely negative correlation, with feeding being more predominant in the middle of the day and play in the early morning and again between 1400 hours and 1900 hours. A similar activity cycle, with play periods predominant in the mornings and evenings, was reported by Cooper (op. cit.). There are four distinct peaks for play, the highest between 1800 hours and 1900 hours when it constitutes 78% of all activities.

Ewer (1968) notes that play in young mammals occurs as a regular part of their daily routine. She mentions as the most frequent times when play occurs in the daily activity cycle the time after feeding and the time following arousal after a period of sleep. As feeding in cubs occurs more or less irregularly throughout the day, play activity cannot be related to feeding times. However, the evening peak of play activity fits into the general rule stated by Ewer, as it follows the daily rest period.

RESTING POSITIONS; DEGREES OF ALERTNESS

The positions of rest have been tabulated into four categories with a gradient of alertness (Table III–1). Number 4 is the most alert and 1 the most relaxed position. These degrees were determined by considering the amount of movement necessary to get into a fully alert position with head raised above the ground, able to look into the distance and ready to move (Leyhausen, 1956 a).

Table III–1
Positions of rest

		Total minutes observed: 707 Positions male				Total minutes observed: 2366 Positions female		
	1	2	3	4	1	2	3	4
Minutes	10	297	58	342	171	1573	87	535
Per cent of total	1·4	42·0	8·2	48·4	7·2	66·5	3·7	22·6
		43·4		56·6		73·7		26·3

The four positions are as follows:
(1) Lying on back with legs relaxed;
(2) Lying on side with legs stretched out:
(3) Lying on haunches with head resting on forepaws;
(4) Lying on haunches with head erect.

In positions (3) and (4) Pantherinae keep their forepaws always in front of their chest, even though they may turn one of the forepaws inward at the wrist. In contrast many of the Felinae characteristically fold both forefeet under their body (Leyhausen, 1956 b).

Sitting was not considered a resting position in this study. It usually occurred for short periods only, as a transition from one of the above positions to standing or walking, or as an alert position between two periods of resting.

When the relative times spent in any of these four positions are examined for the male and for the females, the male appears to be on the whole more alert than the females. The total percentage for positions (1) and (2), the two most relaxed, is 43·4% for the male, as against 73·7% for the lionesses. This may be due to the fact that the females are seldom resting alone; thus any one of them can 'afford' to be more relaxed as the alert position of one of them will be enough to alert the others. The male, even when found associated with a pride, was, as a rule, resting apart.

Position 1 was thought by Kühme (1966) to be possibly a protective measure against too strong irradiation of the spine.

Another explanation for the function of this posture is that it aids in heat loss from the body areas not covered by hair. Dogs and cats adapted to temperate climates exhibit this behaviour frequently when moved to tropical countries. (Prof. D. Robertshaw, pers. com.)

NIGHT ACTIVITIES

An attempt was made to follow the animals four or five nights every month during the full moon. Due to difficulties in the terrain and, frequently, poor visibility, this was not always possible and contact was often lost during the night. Records of nights when contact was kept with the lions at least until 2300 hours, or for a minimum of four hours, were analysed. Observations made on forty-two nights are examined.

The females of the pride started to move away from the daytime positions between 1800 and 1900 hours. In 60% of observed cases the pride started moving between 1830 and 1900 hours.

Unless on a kill, the pride moved intermittently throughout the night. Larger cubs accompanied the females for part of these wanderings but were left behind quite often for periods of up to five or six hours, to be collected again later by the lioness. The male was often seen accompanying a pride, following the females at a distance.

The purpose of the lions' nightly wanderings is the search for prey. Like the tigers at Kanha, whose method of hunting consisted of walking through their range in search of prey (Schaller, 1967), the lion's hunting starts with a search not directed at any particular animal (Kruuk & Turner, 1967). All hunting activity observed by day was triggered off by a prey animal, whilst at night the lions moved off spontaneously after each of their respective rest periods. The proposition that these nightly walks were actually motivated by the search for food finds additional support in the fact that in most instances, when prides were found feeding, the condition of the carcass indicated that the animals must have been killed between dusk the previous day and about 0800 hours that morning. In this restriction of the active search for prey to the night, the Nairobi lions resemble those at Ngorongoro (Estes, 1967), but differ from the lions in the Serengeti which hunt by day at certain times of the year (Schaller, pers. com.).

Hunting behaviour will be discussed in more detail in Chapter VII.

DISTANCES COVERED BY NIGHT

Distances covered between 1900 and 0700 hours ranged from less than 0·5 km. to a maximum of 11·2 km., speed of travel from 0·03 to 2·5 km./hr. Tigers at Kanha are believed to travel 16–32 km. during a night of unsuccessful hunting (Schaller, 1967). The figures indicate that, as a rule, the prides covered less than one kilometre per hour (Table III–2), with an average of 0·7 km./hr.

Unfortunately, for various reasons, the data on night activities do not lend themselves to as thorough an analysis as was hoped. It proved more difficult than anticipated to follow one particular pride or individual for a number of consecutive nights, a condition necessary to get a definite indication of the changing pattern of night activities in relation to hunting success or otherwise.

Table III–2
Distances travelled
by night

Distance travelled km.	No. of nights	Per cent of total
0 to 0·5	9	21·4
0·5 to 5	23	54·8
5 to 10	7	16·7
over 10	3	7·1
Total 42		100·0

Range: 0 to 11·2 km.

Speed in km./hr. Max. Min.	Average of 33 obs.
2·50 0·03	0·67

It was thus not possible to correlate in a definite way hunting success with distance travelled. The conclusions drawn are therefore only tentative.

Table III–2 shows all the nights observed (within the limitations stated above) grouped according to distances travelled. In nine cases (21·4% of total) there was less than 0·5 km. displacement. In five of these cases the pride was on a kill and did not move at all throughout the night. In one case the

Fig. III–5
Day range and
night routes—
Romola pride—
Per. I

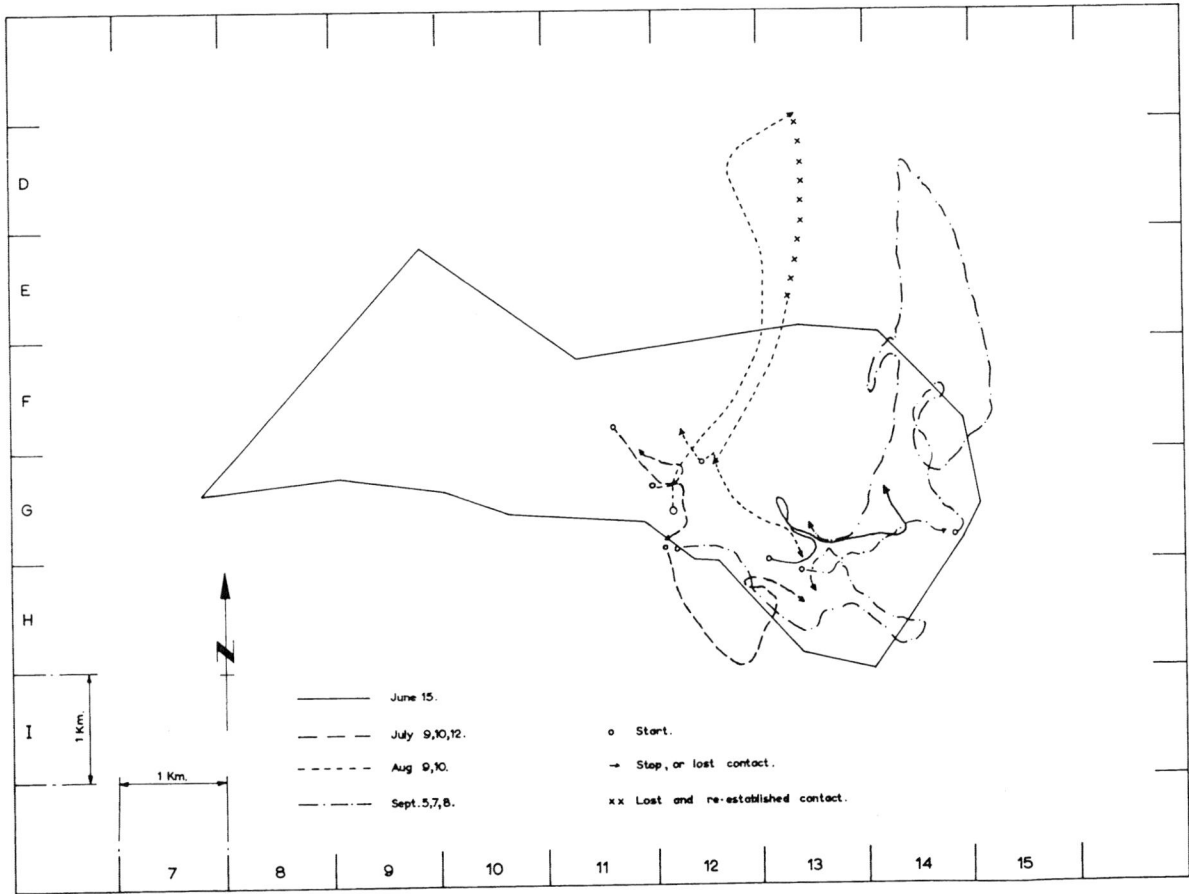

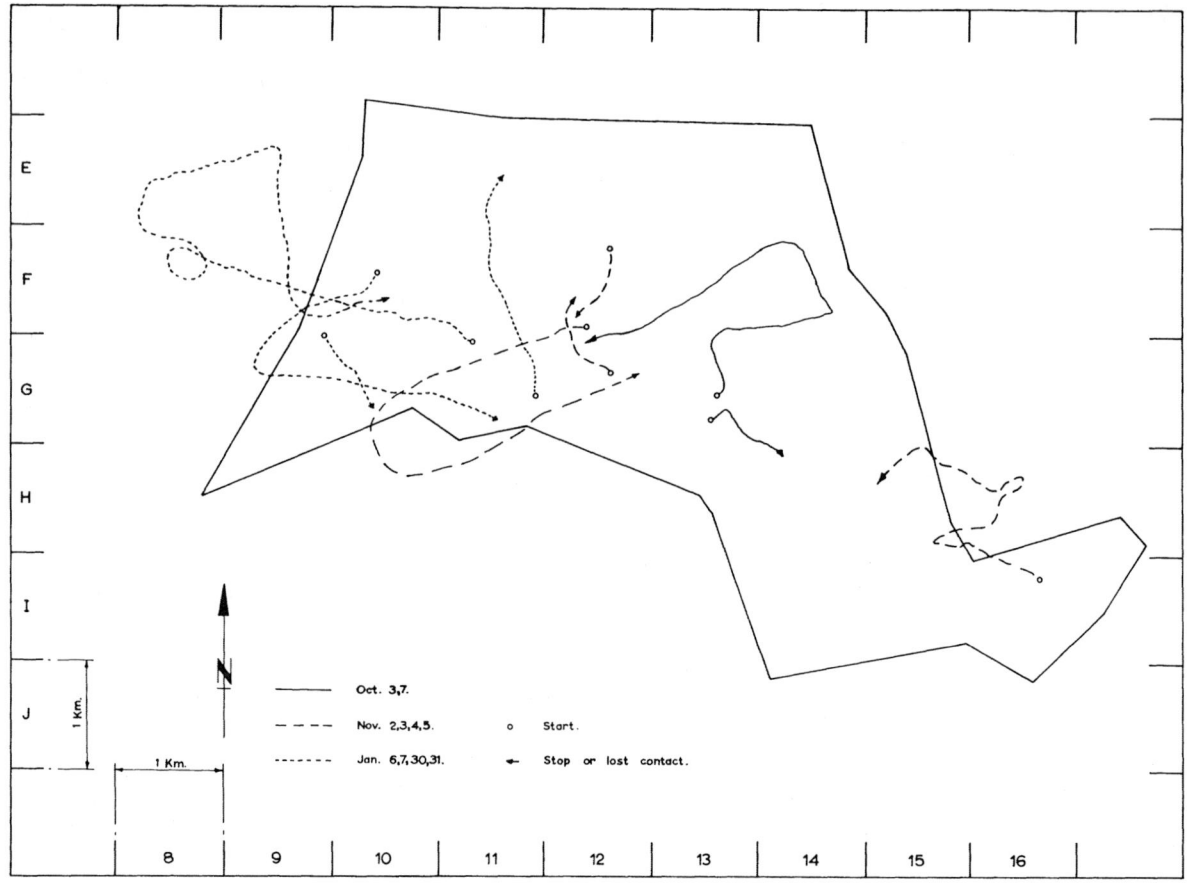

Fig. III–6
Day range and
night routes—
Romola pride—
Per. II

pride was on a kill and left the stripped carcase at 0400 hours at the approach
of a herd of buffalo. In a second, two animals were mating and in another a
pride was observed only between 0230 and 0750 hours during which time
they did not move. Only on one occasion when a pride was actually observed
all through the night (1800 to 0650 hours), was the displacement less than
0·5 km. with the animals neither on a kill nor mating. In over 50% of all cases
the prides moved between 0·5 and 5 km., in 16·6% between 5 and 10 km. and
in 7·1% over 10 km.

The figures do not lend support to the findings of Wright (1960) who saw a
positive relationship between the age of cubs in a pride and the distances
travelled by the pride during the night. Wright gives the maximum distance
covered by a pride with four-month old cubs as 5·4 km. (6000 yd.) and shows
the maximum figure of 9 km. (10 000 yd.) only for a pride with cubs at least
eighteen months old. However, Romola's pride with four and six month old
cubs covered 10·4 km. in one night and with cubs nine and eleven months old
covered 11·2 km. This may be due to the fact that Wright's data were based
on the distances between daily resting positions and thus do not give a true
picture of the distances that may have been covered during the night. It was
observed on several occasions that a pride's resting points on two consecutive
days were less than 500 m. apart, whilst the pride had covered several
kilometres during the intervening night.

If, however, we disregard distances travelled by night and only consider
Wright's data as signifying that the pride's daytime range was more restricted
while the cubs were young, then his observations are supported by the findings
in this study (see Chapter II).

Only on two occasions was there any indication that a hungry pride might

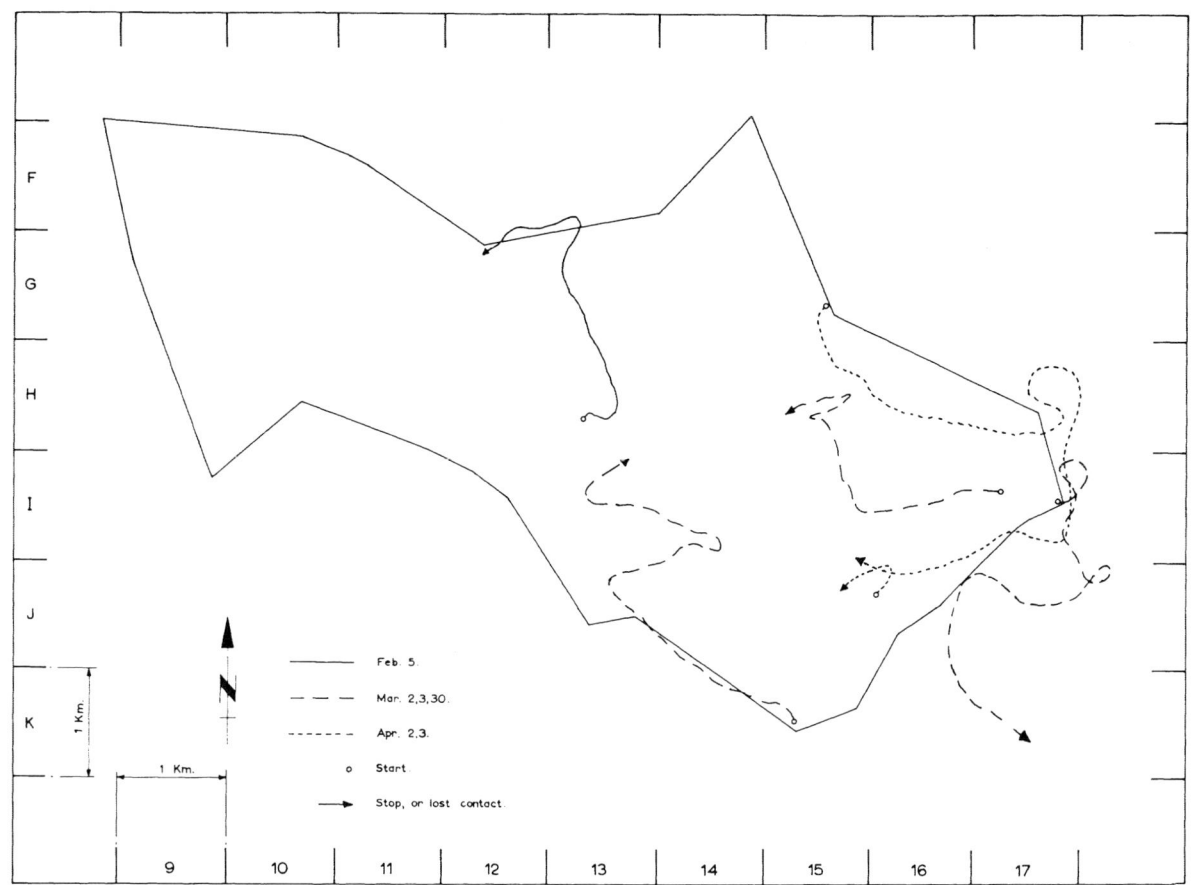

Fig. III–7
Day range and
night routes—
Romola pride—
Per. III

travel further, or faster. Romola's pride was seen feeding on 6th September. On the night of 7th September the pride travelled only 2 km. and made no kill. On the following night, 8th September, they travelled 10·4 km. Again, Romola's pride was last seen feeding on 28th February; two days later, 2nd March, they travelled 6 km. during the night at a speed of 0·75 km./hr.; on the night of 3rd March, they again travelled 6 km. but their speed was 1·71 km./hr.; they made a greater effort to kill, though unsuccessfully, on the second night.

RANGE UTILIZATION BY NIGHT

An examination of the routes taken by the Romola pride (on which the greatest number of observations is available) shows that the area utilized by night did not coincide with the daytime range for the pride for that particular period of the year. Fig. III–5, 6 and 7 show the day ranges of this pride for the three periods and the night routes taken by them during these times. The area utilized by night extended well beyond the boundaries of the day range. A comparison of these routes with the areas of *Acacia drepanolobium* on Fig. II–8, shows that many routes extended deep into the *Acacia* scrub area.

The conclusion that the *Acacia drepanolobium* areas are much more widely utilized by night than by day finds additional support in the fact that between August 1968 and June 1969 the Romola pride was seen twelve times between 0645 and 1000 hours (apart from the night observations discussed above) coming from an *Acacia* area, walking to a resting place outside of the *Acacia* scrub. During the same period they were observed on three evenings between 1830 and 1900 hours entering such a thorn scrub area.

IV Individual Activities

LOCOMOTION

Locomotion in felids includes walking, trotting, galloping and stalking, this latter a gait peculiar to predators, and achieving its fullest development in this family (Leyhausen, 1956 a); it will be discussed in the Chapter on Hunting (Chapter VII). Walking may be of two types: in one type diagonally opposed legs alternately support the body; in the other a bilateral phase is accentuated, with two legs on the same side moving either together or with a slight phase shift; this is known as pacing. Felids may use either type, changing from one to the other in an irregular pattern (Leyhausen, op. cit.); only the Pantherinae among them seem to prefer pacing. This was noted in lions by Schneider (1940, cited in Leyhausen, 1956) and was confirmed by observations during this study. In most cases when the animals were observed walking, they adopted a more or less pure pacing walk, as is illustrated by the pictures on Plate 5 a–d. Pacing is considered an energy-saving type of locomotion in dogs and horses (Scheunert-Trautmann, 1957) and could probably be regarded in the same way for the lion.

The Nairobi Park lions frequently walk along roads and man-made tracks; this habit is shared by the lions of Kruger Park, where they are said to hunt from the roads in the wet season (Pienaar, 1968). When the lions used the main roads, their footprints were clearly visible in the early mornings on the loose surface.

SPEED

Speed was calculated in two different ways:
(1) Strides per minute were counted and length of stride measured from footprints on roads, then the two figures multiplied to give an approximate distance covered per unit time;
(2) animals were followed, the distance measured and times checked with a stopwatch.

Method 1: Distances between two successive imprints of the same foot, one stride, ranged from 96·5 cm. to 132 cm. A mean of 114·2 cm was taken and multiplied by the mean strides per minute (fifty) for male and female animals (Table IV–1). These figures give an average speed of approximately *3·4 km./hr.*

Table IV–1

Sex	No. of obs.	Walking Strides per minute		Trotting Strides per minute	
		range	x	range	x
Male	20	44–56	48·4	—	—
Female	25	46–56	51·4	48–64	59·5

Method 2

(i) 23rd January 1969; pride of two females and five cubs (eight and ten months old) walked from 0735 to 0822 hours with six min. stop for drinking: forty-one min. total walking time.
Distance: 2 km. *Speed: 2·9 km./hr.*

(ii) 9th June 1969; pride of two females and five cubs (thirteen and fifteen months old) walked from 0700 to 0750 hours at steady medium speed.
Distance: 2 km. *Speed: 2·4 km./hr.*

(iii) 9th July 1969; one female leaving kill and walking to cubs from 0820 to 0900 hours with short stops totalling one min. Total walking time: thirty-nine min.
Distance: 2·4 km. *Speed: 3·7 km./hr.*

(iv) 18th April 1969; one lioness and three five-month old cubs from 0710 to 0755 hours steady fast walk and occasional trot without stopping: forty-five min.
Distance: 3·6 km. *Speed: 4·8 km./hr.*

The mean speed obtained from these four figures is *3·5 km./hr.*, which compares well with the mean obtained by the first method, *3·4 km./hr.*, but is lower than speeds quoted in the literature: 4 km./hr. (Labuschagne and van der Merwe, (1963) and 6·4 km./hr. (Guggisberg, 1961) for lions and 4–4·8 km./hr. for tigers (Schaller, 1967).

No attempt was made to calculate speed for trotting or galloping though stride per minute for trotting is shown on Table IV–1. Trotting was observed infrequently and galloping only on rare occasions, when pursuing prey or during high intensity agonistic behaviour, when pursuing another lion. (Chapter V).

PATTERN OF WALKING

Lions were seen to walk continuously without stopping only under special circumstances. Usually many stops were made; some for rest, with the animal lying down, and many more during the actual walking periods, some for elimination or marking, and others without any such activity.

Table IV–2 shows relative duration of time spent stationary and in actual locomotion during walking periods. According to this example, which is typical of walking by the male, one fourth of the time was spent stationary although not resting.

A sustained walk with few stops of very short duration is often seen when a female is returning to her cubs, or when leading them to a kill. These walks give the impression of a purposeful activity without hesitation or uncertainty. On one such occasion Chryse was returning to her cubs; in this case less than 3% of the time was spent stationary. The stops were not only shorter than in the previous example, but also fewer: only about one every 480 seconds as against one every 138 seconds for the male.

Table IV–2

Date Time indiv.	Per. of rest		Per. of walking		Stops during walk		Actual locomotion in seconds	Percentage of actual locomotion during walking periods
	No.	Total duration in sec.	No.	Total duration in sec.	No.	Total duration in sec.		
16th Oct. 1968 0705 0747 Scarface	2	1140	2	1380	10	345	1035	75
9th July 1969 0800 0900 Chryse	—	—	6	2400	5	60	2340	97·5

Another example of such sustained walking occurred on the night of 5th March 1969, when Lassie led her cubs to a kill. This period of walking resulted in the highest km./hr. figure of any night, showing the same type of purposeful uninterrupted locomotion towards a specific goal.

Apart from the two examples of steady walking listed above, fairly uninterrupted periods of walking were sometimes observed in the morning between about 0700 and 0900 hours when the prides were walking to their chosen resting place for the day. (See examples (i) and (ii) in method 2 for speed calculation.)

At night the pattern of locomotion usually showed alternate walking and rest periods, both of very variable duration. The walking phases were usually interrupted with many stops, during which the animals, if not eliminating or marking, stood still looking into the distance.

For a typical pattern of walking by night we may examine the record of the night of 29th–30th July 1969, when contact was kept with the Romola pride and Scarface from 1900 until 0038 hours when the animals crossed the Embakasi river and left the Park.

When resting, the lionesses and cubs were typically 30–100 m. away from the male; when they moved away, Scarface followed them, starting out between five and ten min. later When walking, the distance between the pride and the male varied, contact was sometimes lost for short periods, only to be re-established again. A record of the number and duration of resting, walking and stopping periods was kept for the male during this time and is shown in Table IV–3. Of the fourty-five stops thirteen were made for marking, seven for elimination and twenty-five without any such activity. During several of the latter stops Scarface was sniffing extensively the area where the lionesses had just been lying; very often he then lay down on exactly the same spot.

Table IV–3

Date indiv. Time	Per. of rest			Per. of walking			Stops during walks		
	No.	Duration in sec.	Range	No.	Duration in sec.	Range	No.	Duration in sec.	Range
29th July 1969 Scarface 19 –0038	15		30–762	16		2–186	45		2–60

TREE CLIMBING

In certain areas, such as Queen Elizabeth Park in Uganda and Lake Manyara Park in Tanzania, lions have become famous for their habit of climbing and resting in trees. It has been suggested that the habit may have started as a means of escape from elephant or buffalo, and several such incidents were observed and reported (Cullen, 1969; Makacha and Schaller, 1969). During this study lionesses were occasionally seen to jump up into the lower branches of trees—usually *Acacia kirkii* or other *Acacia* spp—however, they were never observed to rest in that position. They jumped up into the first fork then down again after a few minutes; sometimes this was preceded by claw stretching (see page 42). Tree climbing appeared to be a playful exercise.

Cubs are very proficient at climbing and from the time they were first seen (believed to be about two months old) they very frequently played among the branches of trees (Plate 20 a). Lions descend head first; cubs were sometimes hesitant and made several attempts before descending.

While in Queen Elizabeth and Lake Manyara Parks lions of any age or sex share the habit of resting in trees, the adult male in Nairobi Park was never seen to climb a tree. This, together with the fact that he was never seen to play in any other context, supports the impression that, while in other populations tree climbing may have a functional role, in Nairobi Park it is done in play only.

HEALTH AND SELF-CARE

Injuries, diseases and parasites

Injuries could be sustained under the following conditions:
(1) Intraspecific fighting
(2) Hunting
(3) Accident (playing, etc.)

Table IV–4 shows the number of instances when wounds or scars (Plate 6 a) or other disabilities (i.e. limping), which seemed to be attributable to injury rather than to disease, were noted. It shows that the highest number of injuries were experienced by the young female, Chryse. This may be partly due to her inexperience in hunting (category 2) and partly to causes falling into category 1) as this lioness was the lowest in the rank order in her pride and was often the object of aggression by the other two females (see Chapter V).

The lowest number of injuries in this pride was sustained by Romola, the oldest, hence the most experienced in hunting and the least subject to aggression.

The second highest number is associated with Lassie, possibly due to the fact that after February 1969 she was not associated with any other female; thus she may have been subject to greater risks than if she were hunting in the company of others.

The male, Scarface, was seen only twice with fresh injuries, the first time at the beginning of this study, in February 1968, when he had been in the Park only about three months. Many fairly fresh scars, appearing to be scratches or bites covered his nose, upper lip and paws. They did not seem injuries inflicted by a prey animal, but were presumably the result of intraspecific fighting. This may indicate that the animal was involved in agonistic encounters with the then resident male or males. As Spiv, one of the previously established males, was last seen in the Park at about this time, it may be assumed that Scarface proved superior and displaced him. During the whole period of observation after that Scarface sustained only one injury: a round wound on his chest, possibly made by the horns of an ungulate. As there were no other adult males in the Park for most of this study period to challenge his supremacy and as most of the hunting seemed to be done by the females, he was less subject to injuries than any other adult animal.

Table IV–4

Individual	Sex	No. of injuries
Scarface	m	2
Romola	f	1
Patricia	f	4
Chryse	f	8
Lassie	f	6
Bertha	f	4
Misty	f	1
Nike	f	3
Victor	m (cub)	1
Karen	f ,,	1
Carla	f ,,	2
David	m ,,	1
Pamela	f ,,	1
	Total	35

Lumps on or near the spine were seen on two occasions. An approximately 6–8 cm. long subcutaneous swelling over the lumbar section of the spine was observed when Scarface was first seen in March 1968; the condition still existed at the end of the study although the swelling had decreased somewhat. The lioness Mathilde had a similar lump about 8–10 cm. long, behind the shoulder, close to but not over the spine. The condition was still present when the lioness disappeared from the Park in July 1968.

The only other chronic condition observed was an eye disorder in Lassie, where intermittent discharge from the right eye caused a permanent bare patch underneath it.

Other evidence of impaired health were occasional symptoms similar to a 'cold' in man: sneezing, coughing and wheezing, with discharge from the nose. These conditions were never observed for more than two days running. On 12th December 1968, Scarface was seen sneezing 7 times at an average of once every fifteen min., the next day he was only heard sneezing three times at sixty min. intervals.

Parasites: The lions were only occasionally seen to harbour a considerable number of ticks (Plate 6 b). The largest number of ticks counted on one animal was sixty-nine; these were attached to the shoulder region and the right thigh. The left thigh could not be seen so the total number was probably even higher. Most ticks were seen during February, July, August, October and November.

Internal parasites identified in faecal samples of both cubs and adults were: *Ancylostoma* sp (hookworm), *Ascaroidea* and *Coccidia*. Eggs of *Taenia* sp. as well as eggs of an unidentified fluke were also noted. Dr. J. Cheney of the Veterinary Parasitology Department, University of Nairobi, who kindly identified these parasites, stated that he did not find any reference in the literature to flukes in lions.

Body functions

Breathing: Normal breathing observed in the resting animals ranged between

53 and 66/min. with a mean of 57·3/min. After exercise, such as dragging a carcase or killing a prey animal, the lions panted heavily and their rate of breathing increased to 120–148/min. with a mean of 134·5. This is of the same order as breathing in cheetah after a hunt: 120–150/min. (Schaller, 1968).

Elimination: When an animal is about to defecate it will usually move a few metres away from its position whether resting or walking. A resting lion will move away from its companions, then after defecating will return to its previous position. When a lion was walking on the road, it usually moved onto the grassy surface along the road to defecate, then returned to the road and continued walking. This may be in connection with the tendency of some felids (Leyhausen, 1956 a) to prefer loose substrate for defecation as these animals dig a shallow depression before defecating, then cover up their faeces. Although no such digging, nor covering up, nor any type of scraping motion was ever observed in lions in connection with defecating, the preference for loose substrate may be a remnant of such behaviour.

Defecation took place in well defined periods of the day. 52·7% of a total of fifty-five observations occurred between 1800 and 1900 hours. The synchronization of this activity may be due to two factors: most animals in a pride feed at the same time, and thus elimination would be expected to occur in most of them after a similar lapse of time; the second factor is the seemingly 'contagious' nature of defecation, similar to yawning in man and many other mammals, including lions (Walther, 1963).

Urinating also seems to have a 'contagious' quality. However, in temporal distribution throughout the day it is more widely spread than defecating. Only 37·1% of the total fell between the hours 1800 to 1900; the hour between 1300 and 1400 was totally free of this activity, and during the period of 1400 to 1600 hours there were two instances only out of a total of 167. The rest were fairly evenly spread over the remainder of the day.

Urine may be expelled as a steady stream (Plate 6 d) or in squirts; in the latter case the excretion from the anal gland is mixed with the urine resulting in a strongly smelling scent mark (Schaller, pers. com.).

Scraping with the hindfeet while urinating is a way of marking the living space (Schenkel, 1966) and will be discussed in Chapter V, together with the purely marking function of squirting urine onto raised objects (Fiedler, 1957). The usual body position for urinating is a crouch with extended tail (Plate 6 c). The lowering of the hindquarters is more pronounced in females than males. The face takes on a concentrated expression, similar in both eliminating activities, described by Leyhausen (1956 a) as 'turned inwards'.

For the purpose of urinating the lions do not necessarily seek a position removed from the other members of the pride; they may just rise, urinate on the spot in the middle of their resting companions and lie down again without any displacement. In some instances an animal urinated onto the flank of a sleeping companion without any reaction from the latter.

Self-care and comfort activities

Grooming, or the licking of various parts of the body with long strokes of the tongue along the fur, is the most frequent type of self-care (Plate 7 c). This type of grooming was observed to occur to some extent at most hours as an individual action. However, at certain times of the day it tended to be part of a grooming session, when all or most members of the pride engaged simultaneously in some type of grooming activity, either self-grooming or social grooming. These periods occurred between 0730 and 0930 hours immediately after the pride settled down for their daytime rest, again between 1100 and 1300 hours and in the evening, from 1700 until about 1830 hours, after the animals emerged from their daytime shelter (see Fig. III–1).

According to Leyhausen (1956 b), the Pantherinae as a whole engage in self-grooming to a lesser extent than the Felinae and they practically never groom those parts that the Felinae groom only rarely, i.e. flanks, back, belly and tail. In fact, all these except the tail were groomed by lions, although rarely. The most frequently groomed parts, by far, are the forepaws, which are usually subject to a thorough licking after meals but even at other times are cleaned more often than any other parts of the body. Table IV–5 shows parts of the body groomed in a total of 266 observations: the most favoured parts are front paws, chest and the mane on the chest. Shoulders, genitalia, flank and face were given less attention.

Table IV–5
Parts of body
groomed

Parts groomed	Males over 1 year old		Females and cubs	
	No. of observations	Per cent	No. of observations	Per cent
Front paws	17	34·7	102	47·0
Chest mane	22	45·0	—	—
Chest	—	—	40	18·4
Face	1	2·0	7	3·2
Shoulder	—	—	5	2·3
Genitalia	1	2·0	12	5·5
Flank	—	—	26	12·0
Other	8	16·3	25	11·5
Totals	49	100·0	217	99·9

All animals groomed their front paws for about half the time of their total self-grooming activity. In males over one year old, when the mane starts to grow, the front part of the mane was the most frequently cleaned area with the front paws second in importance (Plate 7 a and c).

The very characteristic 'face washing' of cats (Leyhausen, 1956) was rarely observed in lions. This consists of the animal first licking the inside of its front paw, then stroking along the side of the face with a downward movement of the paw (Plate 7 b). The male was observed only once to engage in this activity and adult females only rarely. Only one juvenile over eighteen months old was ever seen doing it, younger cubs were never observed to wash their faces. Forbes (1963) also states that a lion cub in his care from four days to eight weeks of age was never seen to wash its face, only groom its front paws and flanks. Possibly in lions social grooming has taken over this function of grooming the head region, while in less social felids this activity must be performed within the framework of self-care. Social grooming will be discussed in Chapter V.

All grooming has a utilitarian function, but at the same time self-grooming can be performed as a displacement activity (Armstrong, 1950) in situations of conflicting motivation or when the attainment of an objective is in some way frustrated. An example of this latter case is a 'cat approaching a window to go out but finding it closed, will often give its face a few brief paw wipes' (Ewer, 1968). As self-grooming is to a certain degree indicative of the amount of stress under which an animal lives (Brain, 1965) the relative occurrence of self-grooming as against social grooming will be examined, as it may be an indication of the animal's relationship with other members of its pride.

Table IV–6 tabulates the number of self-grooming performances as well as the total of grooming performances (self and social) for all members of the Romola pride and Scarface.

Table IV–6
Self-grooming—
Scarface and
Romola pride

Individual	Total No. of grooming performances	Total No. of self-grooming performances	Self-grooming as percentage of total
Scarface	72	72	100·0
Romola	248	93	37·5
Patricia	208	107	51·4
Chryse	212	141	66·5
Calef male cub	79	52	65·8
Karen female cub	59	26	44·1
Carla female cub	66	34	51·5
Victor male cub	32	21	65·6
Pamela female cub	39	29	74·4

Scarface was never seen to groom another animal, nor was the other adult male ever seen to do so. In vervet monkeys Brain (op. cit.) found a similar situation where adult males practically never groom any other animal, whilst females frequently do so.

Among the three adult females the highest percentage of self-grooming was found in the youngest, Chryse.

As we shall see in Chapter V, this lioness seemed to be the least integrated member of the pride, at least during the first six months of the study. She was also on several occasions the object of aggression. Her higher degree of self-grooming may thus be interpreted as an indication that this animal lived under a greater degree of tension than the other two females.

The lioness with the lowest degree of self-grooming was the oldest in the pride, first in rank order (see Chapter V) and presumably socially the most secure.

Among the three older cubs, Calef, Karen and Carla, the male showed a higher degree of self-grooming than his sisters. This may indicate the beginning of the typical male pattern of exclusive self-grooming. There was no such sex difference apparent between the two younger cubs; here the high degree of self-grooming can be explained by the development of this behaviour pattern at an earlier stage than that of social grooming.

Scratching, one of the oldest motor-co-ordinations in vertebrates (Leyhausen, 1956 a), is another form of self-care but was less frequently performed. Parts of the body scratched were: behind the ears, under the chin, the neck, and under the forelegs, in that order of frequency.

Stretching: Felids have two typical patterns of stretching the body (Leyhausen, 1956 b): 1. standing with all four feet close together with the back arched upward, tail hanging; and 2. forepaws far forward close together, back arched inwards, head held horizontally or pointed upward (Plate 8 c). Leyhausen notes that in small cats this posture is often followed by a reverse posture whereby the shoulders stay elevated and the lumbar portion of the back is arched inwards with hindlegs stretched out. This latter posture was never observed in lions, only 1 and 2.

Table IV–7 shows the relative frequency of the two stretching postures in adults and cubs. Type (1) is only one-third as often performed by adults as type (2), while in cubs type (1) is the more frequent pattern.

Leyhausen (op. cit.) found a relationship between the two above types of stretching and the previous position of the animal. No such relationship could be established in the case of lions. Both types were performed equally after sitting, lying or standing, especially after long periods of rest and at the beginning of a period of activity. Stretching was usually followed by a short walk either to a new resting place a few metres distant, or for the purpose of feeding or elimination.

Table IV–7
Patterns of
stretching in adults
and cubs

	Type 1	*Type 2*	*Total*
Adults	14	46	60
Cubs	8	13	21
			—
			81
			—

Clawing along upright objects, such as tree trunks, has often been stated to be 'claw sharpening'. Leyhausen (op. cit.) notes that it is a behaviour found only in felids and some viverrids and is performed in a position similar to the one taken when stretching the forequarters. He describes it as an activity whereby the claws are engaged and then pulled back or downwards against a firm object. This is believed to serve for cleaning and sharpening the claws.

Tree clawing was observed in a total of fifty occasions but in only three (6%) out of these was pulling down evident. The action was rather similar to the stretching of the body with inward arched back, whereby the animal either sits or stands facing a tree, mostly one of the *Acacia* species, (Plate 8 a), bracing its paws against the trunk. The claws were rhythmically extended and pulled back again without the paw moving along the tree trunk. Thus this behaviour in the lion seems to be rather a stretching of muscles of back, forelegs and the retractor muscles of the claws, than an actual claw sharpening and cleaning device.

Tail flicking. The sudden jerking up of the tail beyond the horizontal position was often observed. This is a different motor pattern from swishing of the tail to chase away insects; it is a very sudden jerking movement whereby the tail is jerked up beyond the horizontal and sometimes even beyond the vertical position and may recurve over itself pointing forward (Plate 8 b). It was observed in all age and sex categories under various conditions. All the instances when it occurred can be grouped under two main headings: (1) excitement and (2) conflicting motivation.

(1) *Excitement*: tail flicking was frequently observed during mating. The male jerked up his tail once or several times, usually after dismounting, very rarely before copulation. The female on the other hand performed this movement usually before copulation when it may easily be confused with the female's lifting of the tail as an invitation to copulation (Chapter VI), the difference being that the latter is performed as a lifting movement, up and sideways, and the former as a sudden jerking action straight upwards.

During hunting it was performed by young males at the beginning of a stalk.

During agonistic behaviour: the adult male jerked up his tail immediately after charging a car on one occasion and a lioness on another. A lioness was seen jerking up her tail while crouching and snarling at a visitor stretching out his hand through a car window.

Cubs were observed to jerk up their tails occasionally during play.

(2) *Conflicting motivation, uncertainty*: In this context tail flicking was performed by a socially inferior animal (see Chapter V) in contact with a superior member of the pride. The following are examples of this type of situation:

(a) The young female, Chryse, of Romola's pride approached the lionesses and cubs but was chased away by the cubs. When she again approached cautiously she was greeted and accepted. She jerked up her tail while passing by the two females before settling down in their company.

(b) On one occasion Lassie, and on another Misty, approaching Scarface (superior in rank, see Chapter V), jerked their tail before settling down next to him.

(c) Cubs were seen jerking up their tails immediately after being snarled at by the male, and while moving away from him.

All situations in category (2) seem to contain an element of conflicting motivation between aggression and withdrawal.

While Guggisberg (1961) states that an attack is always preceded by several quick tail jerks in rapid succession, it was never observed to precede an attack in the present study.

Rolling from side to side was seen on several occasions in adults of both sexes, but not in cubs. It was occasionally performed by the male during mating, while in the female it is one of the most frequent elements of mating behaviour, and typical oestrus behaviour in cats (Palen, 1966). In certain other instances it is possibly only a type of self-grooming action similar to rubbing against a post or tree trunk as observed in warthog.

v Social Behaviour

GROUP LIFE; SYNCHRONIZED ACTIVITIES

The lion is the only cat that lives a predominantly social life (Schaller, 1966; Leyhausen, 1956 a). This is still substantially true even though (i) Schaller has shown the tiger to be less intolerant of his conspecifics than was formerly thought to be the case and (ii) Leyhausen (1965 a) found a certain degree of social organization even in solitary cats and other mammals.

In carnivores social living has the advantages of co-operative, and thus more efficient, hunting and caring for the young. Possibly the highest degree of social organization among carnivores has evolved in the canids, especially in the hunting dog, Lycaon, where this co-operation includes amicable and equitable distribution of the food supply.

In lions social organization is looser, perhaps, as Ewer suggests (1968), because in canids it is phylogenetically more ancient and thus more highly developed, or it may be connected with the fact that, due to the lions' more specialized killing technique, group co-operation in hunting and killing is less essential for them than for the canids.

The highly efficient, and, I believe, unique killing technique of the muzzle-hold, to be discussed in more detail in Chapter VII, has probably evolved in the course of group hunting. As it is eminently adapted for killing animals larger than the predator, it may become perfected in the course of evolution to the point where lone killing will become as efficient as any co-operative technique and may then tend to diminish the survival value of group life for lions.

In all social groups certain interactions have evolved which, besides occasionally having utilitarian functions, also serve to strengthen the bond between group members. Some such group actions occur between pairs of animals, such as mutual grooming and greeting, the former with, and the latter without, added utilitarian functions. Others, such as group scratching and group yawning in wild dogs (Kühme, 1965) or group howling in wolves and group roaring in lions (Ewer, 1968; Cooper, 1942) are essentially individual activities though performed simultaneously. These may only serve to synchronize mood in the group or generally strengthen group cohesion.

The essentially individual activities which have a contagious quality and occur in synchronized 'sessions' are defecating, urinating, yawning, self-grooming, tree clawing and roaring. Almost half (40%) of 222 observed cases of urinating and defecating occurred as group activity, here defined as at least two animals performing the act within ten minutes of each other. In the majority of cases (57% of all group cases) the time difference was less than one minute.

The contagious nature of yawning in man is well known. Cats in general yawn often (Leyhausen, 1956) and lions tend to yawn repeatedly and in chorus. Individual repeated yawns occurred up to five times within 12 minutes and in group yawning up to five yawns by two or more individuals were observed within 14 minutes. Cubs tend to yawn much less than adults. All yawning observed (ninety-five individual observations) occurred between 1530 and 1830 hours, indicating that lions, unlike cats, do not usually yawn before sleeping but only after resting, while cats yawn both before and after resting (Leyhausen, op. cit.).

Tree clawing described in Chapter IV also belongs to the group of contagious activities. In only two out of fifty observed cases was an animal seen to engage in this activity alone, in all other instances the first tree clawing was followed within a few minutes by one or more animals clawing the same tree.

Whilst roaring will be discussed in more detail later in this Chapter, the group aspect of this activity must be mentioned here. Communal roaring is specific to lions. As Leyhausen (op. cit.) observed in captive animals, tigers do not react to one another's roaring, but lions invariably do so and mostly join in the vocal display. Thus communal roaring is a regular feature among captive lions (Leyhausen, op. cit.; Cooper, 1942), and may occur as often as every forty-five minutes. Whilst communal roaring does occur in lions in the wild, it is by no means as frequent or regular an occurrence as these observations would suggest. Whilst with captive lions at least one other individual almost invariably joins in the roaring, group roaring was observed only infrequently during this study and was heard almost exclusively between sunset and sunrise.

The most frequent and regular group activities were greeting and grooming sessions. The latter occurred in three distinct periods of the day, between 0730 hours and 0930 hours, between 1100 hours and 1300 hours, and between 1700 hours and 1830 hours. The first period is the time when the prides settle down for their daily rest, the last is towards the end of the rest period just before breaking up for the more active night hours. These sessions consist of animals grooming themselves as well as each other. The morning sessions have a more utilitarian function than the other two periods, as usually most of the animals groom themselves; sometimes this consists of cleaning their paws after feeding. The function of the midday and evening sessions tends to be more social than utilitarian. These periods may last from a few minutes to almost an hour. The midday and evening sessions are usually of longer duration than the morning period.

The evening grooming session is usually accompanied by a greeting ceremony, when the adult females greet each other in turn and are greeted by the cubs. (For description of greeting and further discussion see Amicable Behaviour). A similar grooming and greeting ceremony was observed by Kühme (1966) in the Serengeti lions around 1800 hours. In wild dogs a period of greeting, defecating, urinating and playing precedes their evening hunting excursions. Whilst Estes and Goddard (1967) interpret this as a means of bringing the pack to the necessary pitch of excitement for the hunt, Kühme (1965 b) believes that it serves to reinforce mutual dependence and friendliness.

This latter function probably underlies the grooming and greeting sessions in lions, occurring similarly after the daily rest and before the hunting period.

AMICABLE BEHAVIOUR

Mutual grooming

We have discussed self-grooming in Chapter IV and in the preceding section mentioned group grooming as a synchronized activity when it occurs both as self and as social or mutual grooming. Now we shall consider the social aspects of mutual grooming, its function and its occurrence in relation to the individual relationships within the group.

Amicable social grooming is practised in most felids only within the family (Leyhausen, 1956), but occurs in many social animals between individuals of the same group; it has been described in cattle, zebras, vervet monkeys and baboons, where it plays an important part in the social structure (Schloeth, 1961; Klingel, 1967; Brain, 1965; Barrett in Ewer, 1968; Zuckerman, 1932). It consists of one animal licking another or two licking each other simultaneously, or alternately along or against the lay of the fur. In most social grooming the most frequently groomed parts are head, neck and shoulders (Leyhausen, op. cit.; Schloeth, op. cit.). (Plate 9 a–d).

Whilst in zebras adult females rarely groom each other, but males often groom adult females, (Klingel, op. cit.), in lions, as in baboons and vervet monkeys (Brain, op. cit.; Barrett, op. cit.), it is most frequent among adult females. Adult male lions were never seen to groom any other adult. Grooming between females and cubs was also frequent and will be discussed in Chapter VI.

The utilitarian function of social grooming is indicated by the fact that the most frequently groomed parts are those not accessible to the animal itself. The most frequent type of grooming is licking of the partner's fur, and if this is done against the lay of the hair it serves as a more effective cleaning device than when done along the fur as in self-grooming. It was also occasionally observed that an animal was performing small nibbling bites along the skin of its partner and on occasion seemed to be spitting out something. It is possible that ectoparasites (ticks?) are thus removed.

However, as mentioned earlier, social grooming, as one form of amicable social interaction, also functions to maintain and reinforce the bonds between

Passive	Scarface	Romola	Patricia	Chryse	Lassie	Total	Per cent
				Active			
Scarface	—	—	—	1	—	1	0·4
Romola	—	—	48	51	—	99	41·5
Patricia	—	58	—	31	—	89	37·2
Chryse	—	28	20	—	1	49	20·5
Lassie	—	—	—	1	—	1	0·4
Total	—	86	68	84	1	239	100·0
Per cent		36·0	28·5	35·1	0·4		100·0

Table V–1
Grooming between adults of Romola pride, Scarface and Lassie

Total grooming interactions, active and passive	
Romola	185
Patricia	157
Chryse	133
Scarface	1
Lassie	1

the individual members of a group. As such it may serve to indicate the relationships between the individual members within the social organization.

On the basis of mutual grooming I shall try to establish a tentative pattern of social relationships among the three adults of the Romola pride together with Scarface and Lassie, who were occasionally associated with that pride. Then I shall examine these results in the light of further evidence as provided by other types of social interaction.

Table V–1 shows active grooming interactions between these five animals. As mentioned earlier, Scarface was not seen to groom any other adult, nor was the other adult male, Kihara, observed during the early part of the study, ever seen to do so. Scarface was groomed only once, by Chryse, the youngest of the females.

The occasional licking of the partner, performed by the male during mating, cannot be considered social grooming as here defined (more than five strokes of the tongue required). It should rather be regarded as an abbreviated, symbolic form of grooming, incorporated into the pattern of courtship and mating behaviour and having acquired a signal function. (See Chapter VI.)

Whilst both Romola and Patricia were the active partners in less than 50% of their grooming interactions (46% and 43% respectively), Chryse was the active partner in 63% of her's.

On the basis of mutual grooming Romola and Patricia were closer to each other than either was to Chryse, as they groomed Chryse less often than each other, while Chryse was somewhat closer to Romola than Patricia, but closer to the outside members, Scarface and Lassie, than were either of the two other lionesses.

One more aspect of this deserves attention. Table V–2 breaks down grooming activity between Romola, Patricia and Chryse into two periods: June to December 1968, and January to June 1969. Whilst the total involvement in grooming activities shows the same ordering for both periods, with Romola highest and Chryse lowest, there is a slight change in the two periods in the relative percentage of active to passive participation in each of the three. Whilst Romola became more active and received less grooming attention, Patricia showed the opposite trend. Chryse, on the other hand, shows an increase in the total amount of her participation in grooming interactions, and all of this increase is due to her being groomed more often than earlier.

Table V–2
Grooming between three females of Romola pride

Passive	Active R	1968 P	C	No.	Total Per cent
R	—	35	37	72	45·0
P	39	—	19	58	36·2
C	15	15	—	30	18·8
Total number	54	50	56	160	100·0
Per cent	33·8	31·2	35·0	—	100·0

Passive	Active R	1969 P	C	No.	Total Per cent
R	—	13	14	27	35·5
P	19	—	12	31	40·8
C	13	5	—	18	23·7
Total number	32	18	26	76	100·0
Per cent	42·1	23·7	34·2	—	100·0

Greeting (headrubbing)

When two lions from the same pride meet, one bends its head towards the other's head or neck, touches with its forehead or cheek, possibly rubs its head along the other's head, or, more probably, under its chin. In the most complete form of greeting the animal may press its whole body along the other's chin; in its most abbreviated form, it may consist of no more than a slight bending of the head towards the other without even touching. When both partners bend their heads towards each other they may touch heads briefly before separating (Plate 10 a–c). This behaviour, which will be called 'headrubbing' hereafter, is similar to the greeting in cats called 'Köpfchengeben' by Leyhausen (1956), meaning 'presenting of the head'. Treading and presentation of the anal region described by the same author as part of the cat's greeting ceremony, was never observed in lions. Tail lifting into the vertical position, also mentioned by Leyhausen (op. cit.), was observed occasionally in a high-intensity greeting ceremony after a prolonged absence or when several members of the pride milled about each other with every one greeting every other amidst much vocalization. Plate 10 d shows such an occasion when Chryse rejoined the rest of the pride and was greeted by the other two lionesses and the cubs.

Leyhausen (op. cit.) states that this greeting ceremony is performed by a lower ranking animal when meeting a higher ranking one or by a female towards a male.

No male was seen to greet any other individual in this manner, nor were adults seen performing this greeting ceremony towards cubs, whilst cubs rubbed heads to adult females very frequently. The lionesses usually responded by grooming the cubs, even if only perfunctorily, with one or two strokes of the tongue.

Headrubbing between various members of the Romola pride, Scarface and Lassie, was tabulated in Table V–3. The first animal to initiate the ceremony was designated as the active partner; in cases where mutual headrubbing was observed both partners were considered as active partners.

Essentially the same picture emerges as that from Table V–1 showing instances of mutual grooming. Neither Scarface nor the other adult male was ever seen to greet another lion by headrubbing. Females were seen greeting a male in this manner, but here it is of interest to note that whilst Romola and Patricia only did so on two and three occasions, respectively, Chryse, the youngest, did so eleven times. In the Athi pride, with three adult females and Sally, a young adult, the youngest female was the one to greet the male most frequently. On only one occasion was one of the older females seen rubbing her head to the male, Kihara, while Sally was seen to do so three times. A similar situation arises over grooming; the only time Scarface was seen being groomed by a lioness, it was done by Chryse, the youngest of the pride.

Table V–3
Headrubbing
(greeting) between
Romola pride
females, Scarface
and Lassie

			Active				
Passive	*Scarface*	*Romola*	*Patricia*	*Chryse*	*Lassie*	*Total*	*Per cent*
Scarface	—	2	3	11	—	16	12·0
Romola	—	—	14	39	—	53	39·8
Patricia	—	13	—	25	—	38	28·6
Chryse	—	6	15	—	3	24	18·1
Lassie	—	—	—	2	—	2	1·5
Total	—	21	32	77	3	133	100·0
Per cent	—	15·8	24·0	57·9	2·3		100·0

Among the three lionesses of the Romola pride, Chryse performed the greeting ceremony more times than the other two females combined, she was on the receiving end of the greeting the least number of times while Romola was greeted most often. Lassie is again shown to have social contact only with Chryse.

Greeting, as grooming, involves close bodily contact. Although lions on the whole are contact animals, resting in close proximity to each other and seemingly seeking bodily contact in various other ways, such as mutual grooming and greeting, adult males do not seem to seek, and indeed, rather try to avoid, such contact. Thus, the adult males not only did not groom or greet others, but were neither groomed nor greeted very frequently by either females or cubs, although, by any other criteria, as will be shown later, they must be considered the highest ranking in their respective group of females and cubs.

The low number of amicable social interactions between the male and the females and cubs was due not only to the fact that the males were only occasionally close enough to them to make such behaviour possible, but to the males' positive avoidance of such close contact. Scarface usually responded in an agonistic manner, by growling or snapping, or in a negative manner, by lifting his chin high so as to avoid contact when another individual approached to greet him.

AGONISTIC BEHAVIOUR

Lorenz (1963) maintains that the more intraspecific aggression occurs in a group, the firmer the individual bonds that are formed; he then goes on to say that where aggression is seasonal, so are the personal attachments. In lions aggression is ever-present, though largely confined to competition over food, and some associations (Romola and Patricia) attest to a very strong personal bond which may last for years (Guggisberg, 1961).

Although agonistic behaviour is frequent among lions, actual fighting is rare, (Schaller, 1969). In carnivores aggression is rendered less damaging to the species by the rarity of actual physical contests rather than by fighting itself becoming ritualized as in herbivores (Ewer, 1968).

Threat and intimidation displays

The function of threat display is to avoid actual physical contests of strength by discouraging the weaker partner from continuing with his current action or direction. Threat often involves a demonstration both of size and of weapons; it can then be called intimidation display (Ewer, op. cit.). Elements of intimidation and threat are listed below and the context in which they occurred is noted.

(i) Snort, grunt, or growl, without turning head in direction of opponent. Usually while feeding, directed towards an animal approaching too closely to head of feeding individual.

(ii) Pushing away with head, front paw or hindleg. Females towards cubs; to prevent feeding or suckling. Once seen from a two-year-old male who tried to push away with his head, two lionesses approaching a kill.

(iii) Blowing audibly through nostrils with cheeks blown up, while turning towards opponent or object of annoyance. Mother discouraging cub from suckling; cub against another cub feeding too close to it; Scarface towards stationary bus. Cheek inflation is known in rodents as a form of threat display; it increases size of head (Ewer, 1968).

(iv) Snarling; retraction of lips baring teeth, especially canines, ears laid back (Plate 11 a and c). May be repeated slowly once or twice, or many times in rapid succession. Head is turned towards opponent; silent or with growling. This type of weapon threat is also used by canids (Ewer, op. cit.). In domestic cats shown only in highest intensity defensive threat (Leyhausen, op. cit.). Seen most often from Scarface, when feeding or when approached by cub while resting.

(v) Similar to above, but accompanied by snapping motion, a rapid opening and closing of jaws; usually with growling. Most frequent while feeding, seen from male, females and cubs.

(vi) Hitting with forepaw. Seen from cubs or females in response to attempts of companions to engage in play. Also by male, females and cubs while feeding.

(vii) Displays involving displacement of whole body:

(a) Standing up, head erect, gaze fixed onto opponent. Lorenz, (cited by Ewer, 1968) regards some human threat postures as related to the erection of neck and shoulder manes present in some primates, but which we no longer possess. It is possible that the 'head high erect' threat posture, seen mostly on the part of the male, is occasionally accentuated with piloerection of the mane. No vocalization.

(b) As above, with few steps in direction of opponent.

(c) Lunging at opponent.

(d) Pursuit and attack.

(b) and (c) may be accompanied by growling. Seen mostly from Scarface while feeding, and occasionally while not feeding, against approaching companion. (d) observed from females against females only.

In only a few instances was an agonistic encounter of intensity (vii) (d) observed. The occurrence of highest intensity involved the three females of the Romola pride who pursued and attacked Lassie. On 5th November 1968, Scarface, Patricia, Romola and five cubs were feeding on a wildebeest. At 1730 hours Lassie and Chryse approached together. While Lassie lay down facing the pride at about 70 m. distance, Chryse joined the group. Both Patricia and Romola were attentive, having noticed the approaching lionesses at a distance of about 100 m. Chryse exchanged greetings with both females and settled down close to them while the other two started feeding. At 1802 hours Scarface approached the feeding lionesses, whereupon they both left the carcase, Romola with a grunt, and together with Chryse walked, then, after a few steps, ran at full speed (gallop) towards Lassie who turned around and ran away. When I caught up with the group a few seconds later, Lassie was about 100 m. from her original position. She seemed to have been thrown over (or had thrown herself on her back?) and was just righting herself into a crouching position with the other three close by and surrounding her. There was much growling from all participants and the defensive lioness showed piloerection along her spine. There was no evidence of actual biting or of hitting with the forepaw, described by Cooper (1942) as the most conspicuous fighting movement in captive lions, but there was much snapping especially by the attacked lioness towards her attackers. After less than a minute Patricia and Romola left her and returned to the kill, while Chryse stayed with Lassie and only rejoined the pride 30 min. later, making several stops on the way looking back towards Lassie, who eventually departed in the opposite direction.

This attack on Lassie, who had, on several other occasions, been associated with the pride without animosity, has all the characteristics of 'redirected aggression' as it seems to have been triggered off by the two females leaving the kill at Scarface's approach. This sequence was comparable to the 'handing down' of punishment mentioned by Zuckerman (1932) when apes, attacked by a higher ranking individual, turn around to attack another animal, below them in the ranking order.

In this case the aggressive drive of the lionesses, elicited by Scarface displacing them from the kill, was at the same time blocked by his superior status (discussed in more detail below) and then 'redirected' towards a substitute object. As Ewer (1968) states, when attack is not preceded by threat it is carried out at great speed and intensity, with no sign of the hesitation or conflicting motivation which is characteristic of threat displays. In this case the attack was launched at great speed without any preliminaries.

On several other occasions, during August, September and October, 1968, Romola and Patricia pursued Chryse, without any apparent provocation from the young lioness. However, in none of these cases was the pursuit followed by a confrontation with growling and snapping, as in the case of the attack on Lassie, nor was Chryse pursued for as great a distance as in the case just mentioned. These pursuits were of a considerably lower intensity.

Defence; submission

Behaviour patterns elicited by another lion's demonstration of aggression are listed below from the least intense, evasion, in order of increasing intensity:
  (i) Stepping aside, evading. Usually as reaction to threats (iv) and (v), occasionally to (i).
 (ii) Jumping (a) aside or (b) straight up. As above. (b) observed once as response of cub to threat display (vii) (c) by Scarface. (i) and (ii) are cited by Schaller (1966) as the most frequent responses of tigers to threat.
(iii) Lifting paw in defence. Seen only once from cub in response to (v).
 (iv) Ducking, body lowered into slight crouch. Usually reaction to (iv), (v) or (vii) (c).
  (v) As (iv) but with head pressed sideways to ground (Plate 11 b). Usually seen from cubs, reaction to (iv), (v) or (vii) (c). (iv) and (v) may be accompanied by vocalization in cubs: screaming or squeaking; usually silent in adult females.
 (vi) Crouch, ears laid back, with or without tail swishing. Seen from all animals as reaction to humans or vehicles; from cubs as reaction to Scarface approaching without any aggressive behaviour. Sometimes combined with moving backwards while maintaining crouching position.
(vii) As (vi) but combined with element (iv) in threat displays, with or without backward locomotion; usually silent. From females and cubs; reaction to humans and vehicles. From cubs as response to approaching male not showing aggression.
(viii) As (vi) combined with growling and snapping (element v) of threat display) and piloerection along spine. Seen only as Lassie's response to lioness's aggression.
 (ix) Rolling onto back; occasionally accompanied in cubs by screaming. In response to (iv) and (vii) (d) of threat elements; seen from cubs and lionesses. Whether claws are retracted or extended during this posture could not be determined but the paws were held limp as Schaller (1966) has seen it in defensive tigresses, where he regarded it as a gesture of submission. Only on one occasion was this posture combined with hitting, when a female lashed out towards the male during food competition. The exposed belly posture was also described for canids (Kühme, 1965 b) but was not mentioned by Ursin (1964) in the domestic cat's repertoire of defensive behaviour. Leyhausen (1956 a) describes it for cats only in its form of active defence, not submission, where the animals use their claws and teeth while on their backs.
   (i) to (vi) may be considered pure defence, (vii) and (viii) as defensive

threat and (ix) as submission in the majority of cases and only rarely as defensive threat.

Leyhausen (op. cit. quoting K. M. Schneider) mentions a posture peculiar to the Pantherinae, where an individual discourages the approach of another by standing with head held high, slightly retracted lips, weaving its head slowly from side to side. This behaviour was never observed during this study.

INTRAGROUP RELATIONSHIPS

Social bonds, individual associations; rank order, leadership

Mutual attachments between individuals are characteristic of certain types of animal groups, where, in contrast to anonymous crowds, individuals are held together by a personal bond. This amicable pattern then develops its own innate drive and the individual needs a companion for its expression (Lorenz, 1963).

The degree of bonding between individuals can be measured by the amount and intensity of the amicable interactions between them. The simplest form of these is physical proximity (Walther, 1963; Kühme, 1965 b) Schloeth, 1961). In the following paragraph we shall examine the relationships of Scarface, the Romola pride and Lassie on the basis of these considerations.

The three females of the Romola pride were usually seen all together; the two mothers, Patricia and Romola, were only seen apart from each other on very few occasions, while they were seen without the youngest, Chryse, on 20% of the occasions when the two lionesses were encountered.

As Walther (op. cit.) observed, the distance between resting animals is a good indication of their relationship to each other. When all three lionesses were resting together, a record was kept of the estimated distance between the closest parts of their bodies. It was noted whether Chryse was further from either of the other two than they were from each other. Records were made on the hour and half hour. Table V–4 shows the results of these observations both for the frequency of Chryse's association with the other two lionesses and the closeness of her position. As there was some indication of a changing relationship between these three animals, the observations were divided into two periods, June–December 1968, and January–June 1969. The results confirmed the impression of a change: the percentage for distances between the resting lionesses are almost exactly reversed in 1969. Whilst in 1968 Chryse

Table V–4
Association of
Chryse with
Romola and
Patricia

	1968		1969	
	No. of obs.	*Per cent*	*No. of obs.*	*Per cent*
Chryse, Romola and Patricia together	82	78·8	60	80·0
Romola and Patricia only	22	21·2	15	20·0
Total	104	100·0	75	100·0
* Chryse further from Romola and Patricia	49	64·5	44	33·3
* Chryse not further from Romola and Patricia	27	35·5	88	66·7
Total	76	100·0	132	100·0

* Half hourly observations.

was furthest away in almost 65% of cases, in 1969 this percentage is reduced to about 33%, whilst the percentage of her association with Patricia and Romola remained remarkably constant at about 80%.

The two older lionesses were seen apart only on five occasions. Patricia was seen without Romola twice when the latter's cubs were less than two months old, and Romola was seen without Patricia on three occasions when Patricia's cubs were less than two months old and not yet introduced to the pride. This is in accord with observations made by Schenkel (1966) of a lioness leaving her pride temporarily for the period of parturition and dividing her time between the newborn cubs and her pride for the first 6–10 weeks.

On the basis of observations discussed in this chapter: (i) agonistic behaviour, (ii) social grooming, (iii) greeting, (iv) association and (v) proximity, the tentative conclusions proposed in the section on mutual grooming seem to be confirmed by data on the other interactions. Chryse's lower status is indicated by the extent of aggressive behaviour directed towards her and by her lack of priority in obtaining food, as well as by her performing the greeting ceremony more often than the other lionesses combined and being greeted the least often. The fact that her relationship to Patricia and Romola was less close than their relationship to each other is also shown by the higher percentage of absences from the pride and by her low number of grooming interactions.

The conclusion that her relationship became closer in 1969 is supported by the following findings: (i) agonistic behaviour towards her of intensity (vii d) was seen only during 1968; (ii) she was groomed more often in 1969 than in 1968; (iii) she was resting closer to the other two lionesses in 1969 than in the previous six months.

Dominance can be defined as possession of priority in access to desired commodities. These may be space, food, mate or companionship (Struhsaker, 1967). The significance of a dominance system lies in its reduction of physical aggression and conservation of energy.

Any group of animals may set up a dominance hierarchy (Scott, 1958, cited in Cole & Schaefer, 1966). Leyhausen (1956 a) describes what he terms 'relative' and 'absolute' rank orders in a free ranging cat population, while in caged animals the balance between these two shifts more and more towards absolute hierarchy with increased crowding. Cole & Schaefer (op. cit.) observed stable hierarchies established in groups of eight cats.

On the basis of priority in feeding and aggressive behaviour, this study confirms the findings of previous observers (Kühme, 1965; Schaller, 1969; Guggisberg, 1961) that the male lion is dominant over the females. The overwhelming majority of threat behaviour was directed by the male towards the females and cubs. In many instances this behaviour was aimed at discouraging another animal from approaching too closely, but the highest intensity of aggressive behaviour was apparent during feeding. An adult male appropriated an animal killed by one or more females in four instances and allowed no member of the pride to approach until he had finished feeding.

On one of these occasions (31st January–1st February 1969) the male guarded the carcase of a Grant's gazelle, on which he was occasionally feeding, against one of the females after the others had departed. Between 2030 and 0645 hours he snarled and growled (threat i) & iv) many times and lunged twenty-five times towards the lioness (threat vii c).

Another typical instance of the male's priority in feeding is seen in the following sequence, when his rank was expressed in 'giving way' by an inferior (Chryse, in this case) rather than by overt aggression on the male's part.

1st October 1968, 0645 hours, the Romola pride was walking in a loose group with Scarface, last in the line, carrying an almost completely stripped carcase of a kongoni. Chryse walked in front of him. The male dropped the carcase and walked on; Chryse returned and pulled at the carcase. Scarface returned to the kill, Chryse left it and walked away. Scarface picked it up, stood still, dropped it, then walked away. Chryse returned to the carcase again, and lay down to feed.

In only one instance was there aggressive reaction to the male's agonistic behaviour in connection with feeding, when a lioness (see this Chapter, Defence and Submission) lashed out against him; in no other case did an adult male meet any resistance when he approached a carcase to start feeding.

In contrast to the findings of Schaller (1969), who found no further rank order among the lionesses, this study indicated the existance of a rank order among the females. In the Romola pride Chryse was the lowest in the hierarchy and Romola the highest, although her dominance over Patricia was not so clear cut.

Whilst Chryse was repeatedly the object of aggression by the other two females, Romola was never observed to be the object of aggression, Patricia very rarely and only in a food competition situation, whilst food competition did not necessarily enter into aggressive behaviour displayed towards Chryse.

Situations similar to the one described above between Chryse and Scarface have often been observed, when Chryse left a carcase at the approach of Patricia or Romola and returned to feed after they had walked away from the kill. When a warthog kill was made by the three females on 3rd March 1969, Chryse started feeding only after all the others had departed, and only scraps were left. Cole & Schaefer (1966) found that when cats had to compete for food, the consistent losers eventually lost interest and did not compete any longer. In the case just mentioned Chryse did not seem to try to start feeding while the others were on the kill, but repeatedly played either with the carcase or one of the cubs.

In two instances when Chryse had caught a prey (once a hare and the other time a steinbuck) Patricia appropriated it before the young lioness could start feeding. The only protest from Chryse was a short growl but she made no attempt to defend the carcase or to take it back.

Zuckerman (1932) suggests that sexual dimorphism may be a factor in male dominance, due to superior physique. In hyaenas females are both larger and dominant over males (Kruuk, 1972), while in all primates, and also the lion, males are both bigger and dominant. From observations during this study it appears that rank order in lion is based on size, as Schaller has suggested for tigers (1966). However, the male's dominance over the female may be due to sex as well as to size, and as the larger females or cubs were also older than the smaller females or cubs, it is possible that age also plays a role in determining rank order.

Whilst dominance clearly belongs to the male, leadership is exercised by the lioness (Schaller, 1969). Even when a male was present, it was always one of the females who initiated any activity and the male followed as the last of the group. When the pride was moving away, the male got to his feet a few minutes after the others and followed the females and cubs at a distance (see Chapter III). On only two occasions was this order reversed, when for about 10 min. the male led a group. Both adult males behaved in the same manner in this respect.

Any one of the females was just as likely as any other to take the lead. Of the three lionesses in Romola's pride the two older ones usually walked close together while Chryse was either in front or behind, however, either one of the three could initiate movement or be in the lead.

As lions are usually only active when hunting, leadership in activities generally means, more than anything, leadership in hunting. Thus the tendency of the males, observed during this study, to lag behind is connected with the fact that most males are not very active in hunting but rely on the lionesses to provide their sustenance (Schaller, op. cit.).

COMMUNICATION: VOCALIZATION AND MARKING

The lion's roar has often been described as the most awe-inspiring sound (Guggisberg, 1961; Schaller, op. cit.) and the frequency of the lion's roaring, in contrast to other great cats, has been noted by Leyhausen (1950). There are, however, a great many other vocalizations of which the lion is capable. Table V–5 lists the various sounds heard, their context and their probable function. Some were heard from females' only, such as the cub-calling sound (16), some are specific to the male (15), or to cubs, but roaring has been heard equally from both males and females, although not from cubs under the age of fifteen months.

The function of the lion's roar has been much discussed; Schenkel (1966) considers it a type of territorial display, whilst Leyhausen (1950) believes that it serves to promote the cohesion of a group.

Lions can roar in any body position (Leyhausen, op. cit; Cooper, 1942), walking, lying or standing. Although they may start in any of these positions, for high intensity roaring the animal will usually stand or lie on its haunches when vigorous movement of the diaphragm can be observed. The roar produced is a series of rumbles, each series being one expiration (Cooper, op. cit.). The mouth is opened wide for a fraction of a second at the beginning of a series then the lips pulled forward and the sound produced with the mouth only slightly opened and head thrust forward in a horizontal position (see plate 12). A series usually lasts less than one minute and may contain up to forty-two rumbles of decreasing intensity. Table V–6 shows duration of series, frequency and number of rumbles (repeats). No significant difference between the male's and the female's roaring could be detected though the male is said to have a lower pitch (Cooper, op. cit.).

The most evident function of roaring, as observed during this study, was to ensure cohesion of the group and to facilitate contact during the night either between members of a pride or between the male and a pride. The following observations illustrate this function of roaring:

(a) 15th June 1968. Scarface roared nine times between 1910 hours and 0145 hours. At 0215 hours Chryse met him and stayed with the male. No more roaring heard until 0650 hours.

(b) 10th July 1968. All three females and cubs of Romola pride were together until 1916 hours when one female left. 2040 hours to 0420 hours roaring heard six times from a distance. 0454 hours, pride started moving, after 150 m. were rejoined by third female. No more roaring heard until 0930 hours.

(c) 8th August, 1968. Scarface, Patricia, Romola and cubs together. Scarface and females roared six times between 0155 hours and 0505 hours when Chryse joined them. No more roaring heard until 0700 hours.

(d) 24th July 1968. Patricia, Romola, Chryse and cubs together. Roars heard from distance four times between 1825 hours and 1843 hours.
All looked and then walked in direction of roar.

(e) 3rd November 1968. Scarface roared five times between 1815 hours and 1900 hours, met two females at 1930 hours. They stayed together, no more roaring was heard and contact was lost at 2245 hours.

(f) 6th January 1969. Patricia, Romola and cubs together. 1855 hours to 1918 hours roaring twice from distance, Patricia and Romola roared three times, then walked in direction of roar. 1940 hours Chryse and at 2100 hours Scarface joined up with them. 2120 hours lost contact.

(g) 31st January 1969. Patricia, Romola and Chryse left cubs at 0140 hours. Roaring heard from distance at 0535 hours, cubs sat up, then ran towards sound, met Patricia and Scarface at 0550 hours.

Whilst establishing contact seems to be an important function of roaring, it is, of course, not the sole purpose of this vocal display. Guggisberg (1961)

Table V–5
Vocalization

No.	Description	Context	Probable emotion	Response	Probable function	Frequency by sex* and age		
						M	F	C
1	Grunt (short) or Snort	M. when approached by F. or cubs; F. discouraging cubs from suckling or feeding; all animals while feeding	Annoyance	Evasion, walking away	Warning, threat (mild)	xxx	xx	x
2	Growl	M., F., cubs, when feeding; M. at approach of F. or cubs	Anger	Evasion, defensive posture; occas. vocal.: (a) defensive (b) growling	Warning, threat	xxx	xx	x
3	Prolonged monotone growl	M. following F. closely during mating	?	None	Warning to conspecifics not to approach F.?	xxx		
4	Blowing through nose	F. against suckling cub; F. or cubs while feeding; directed agst. vehicle by M.	Mild anger	Evasion	Warning	x	x	x
5	Whine	When frustrated in attempt to suckle or feed; when M. charges mother	Annoyance, uneasiness	None	?			xx
6	Moan	When temporarily separated from pride	Slight distress	Cubs emerge from cover; female approaches	keeping pride together		xx	
7	Howl	Response to being hit	Annoyance fear	None	?			x
8	Scream squeak	Response to intense threat; accomp. by defensive posture	Fear	None	May inhibit further aggression			x
9	Yelp	Cub walking about in presence of mother	Slight distress	None	?			x
10	Miaow, high pitched whine	Cubs meeting females; cubs trying to suckle	Desire to be fed	Mother suckles cub or grooms cub, or no response	Expresses desire to be fed, desire for comfort			xxx
11	Roar, sequence of rumbles	M. when walking alone; when pride incomplete; when feeding, or before starting to feed; after high intensity agonistic encounters	?	Roaring; cubs cluster near mother; alert looking in direction of sound; none	Bringing together members of pride; warning to outsiders?	xxx	xxx	
12	Purr	Copulation	?	Mounting by male	?		xxx	

Table V–5 *continued*

No.	Description	Context	Probable emotion	Response	Probable function	Frequency by sex and age		
						M	F	C
13	Grunts (prolonged) or soft moans	When one animal is reunited with pride	Contentment	Rubbing against each other, milling about, high intensity greeting	Reinforces social bond		xxx	xx
14	Grunt (short)	M. while lying down after copulation	Contentment?	None	?	xxx		
15	Long drawn-out single howling roar	Copulation; before or during dismounting	Excitement?	None; cubs, if present, become alert and cluster together	?	xxx		
16	Staccato, hm-hm, hollow sound	When cubs stray too far; shortly before returning mother reaches cubs	?	Cubs become alert; run to mother	Keeping cubs close to mother; warning	xxx		
17	Chirp	Cubs approaching mother	?	None	Wanting to suckle?			x

xxx Frequent.
xx Occasional.
x Rare.

has stated that lions roar while hunting, but they were never heard to do so during this study, though roaring was heard on several occasions after a kill had been made, either before feeding was started or during feeding. It is not quite clear what the function of roaring in this context could be. Schaller (1966) noted that tigers roared often after they made a kill and that they seemed to draw attention to their food supply.

One night Scarface discovered a Grant's gazelle carcase, started feeding on it for a short time, got up, roared, then resumed feeding. Why he should be calling attention to himself in this situation is puzzling, as he usually did not let any other lion approach a kill for some time after he started feeding. Similarly, as will be described later in Chapter VII, when Patricia, Romola and Chryse killed a warthog, Patricia, who was keeping a stranglehold on the animal's neck and thus was the one who most probably killed it, did not start feeding immediately, but walked about, roared, then walked away and returned only an hour later to start feeding.

Roaring also seems to indicate uneasiness or agitation: it is then produced like a soft, moan-like sound rather than a full roar. On 4th August 1968, Patricia and Romola were on a kongoni kill with the five cubs. Many cars were around and there was more than the usual amount of noise as several drivers were revving their engines to get through some very muddy patches. Patricia was walking about moaning-roaring softly, then picked up the carcase and carried it 200 m. away from the cars followed by Romola and the cubs.

Cooper (1942) observed that every fight among the captive lions was followed by a short, but intense, roaring session. In a similar way the high intensity agonistic encounter between Romola, Patricia, Chryse and Lassie, described earlier in this Chapter, was followed by a short roaring display ν the part of the Romola pride and Scarface.

Thus roaring also seems to express tension, as well as the release of a tense situation or one charged with high emotional content. The reaction of cubs to

Table V–6
Roaring: Duration
of series and
frequency of
rumbles

Duration of series sec.	Rumbles No.	Frequency per sec.	Duration of rumbles sec.
90	35	0·4	2·6
40	29	0·7	1·4
56	30	0·5	1·9
50	29	0·6	1·7
39	23	0·6	1·7
42	26	0·6	1·6
41	23	0·6	1·8
41	27	0·7	1·5
45	29	0·6	1·5
40	28	0·7	1·4
47	27	0·6	1·7
35	20	0·6	1·7
43	33	0·8	1·3
48	30	0·6	1·6
39	26	0·7	1·5

	Range	Average
Duration of series	39–90 sec.	46·4 sec.
Rumbles, number	20–35	27·7
Frequency of rumbles	0·4–0·8 per sec.	0·6 per sec.
Duration of rumbles	1·3–2·6 sec.	1·7 sec.

roaring also indicates that it has more than just a contact-seeking significance. In almost every case when the females roared, the cubs immediately stopped whatever they were doing and huddled close to the roaring lionesses. They left them and returned to their respective activities as soon as the roaring stopped.

A soft moaning sound (Table V–5, 6) was heard on two occasions when a lioness temporarily got separated from her pride and tried to re-establish contact.

On 16th November 1968, at 0800 hours the Romola pride was together in a loose group, some resting and Chryse walking about in riverine thicket. At 0814 hours the rest of the pride walked about 80–100 m. along the river and disappeared in the thicket. At 0832 hours when Chryse emerged from the bushes after the others had left, she looked about, moaned softly, then sniffed an *Agave* plant along the track the others had taken. Chryse then called again, whereupon, at 0835 hours, three of the cubs came out to the edge of the thicket and lay down looking towards the young lioness. Chryse then walked towards them and joined the pride.

On another occasion when Chryse got similarly separated from her pride, she used the same moaning call; Romola returned to her from about 30 m away, where the pride lay hidden from view behind a patch of bush, then both joined the others.

As has been suggested by Leyhausen (1956 a) and Schaller (1966), the prominent ear spots on tigers and lions probably have a signalling function. Whilst in the forest-dwelling tiger the spots are white, in the lion they are black and are the only conspicuous marks on a female when seen from the back, especially when lying in tawny grass. They probably aid members of a pride to keep contact during a communal stalk when communication by sound would be undesirable.

Vocal communication serves as a means of contact for animals separated in space, whilst olfactory communication may serve to transmit messages between individuals separated in time.

The function of marking, an olfactory communication, has variously been explained as an action to familiarize the living space (Schenkel, 1966; Baerends, 1956), to ward off intruders (Lorenz, 1963; Kleimann, 1966), to establish property rights on behalf of the group, and to act as a feedback mechanism to control population density (Wynn-Edwards, 1962). Domestic cats and most Pantherinae mark their range with a secretion from their anal glands which they spray onto bushes, trees or other objects raised above ground level. (Leyhausen, 1956 a; Fiedler, 1957). On one occasion Scarface was even seen spraying a sightseeing bus standing close to him.

Besides spraying scent lions also mark by rubbing their heads into the branches of a bush or tree. No scent gland on the lion's head has ever been described and it is possible that in this case only the generalized scent of the animal is rubbed onto the tree. This activity is usually followed by the animal turning around and subsequently spraying scent onto the same area (Plate 13 a, b). This attitude of backing up to an object and squirting upwards with tail raised high (Plate 13 d) is quite distinct from the lion's usual stance while urinating, when the stream is directed downwards in an even flow with the tail relaxed (Plate 6 d). This directional character of the motor pattern is a typical feature of marking activity (Kleimann, op. cit.). That scent spraying and urinating are two distinct activities is shown by the fact that one often succeeds the other, urinating following scent spraying.

Another method of marking is scraping with the hindfeet while urinating (Schenkel, 1966). This will deposit the scent of urine along the animal's trail.

All these methods of marking have been observed in both sexes but are much more frequent in males. Fiedler (1957) made similar observations on tigers in captivity. Cubs up to fifteen months old were never seen to scrape while urinating (Table V–7), nor were they observed to spray scent. Fiedler (op. cit.) noted that in captive lions males start scent squirting only at about two-and-a-half years of age. However, cubs as young as nine months old were seen rubbing their heads into the branches of a bush and once a cub was seen to rub the whole length of its body along a shrub.

Table V–7
Urinating with scraping as percentage of total urinating

	Total obs.	With scraping	With scraping as per cent of total
Male	32	16	50·00
Female	80	27	33·75
Cubs	22	—	—
Total	134		

On one occasion, the male, while walking with occasional short rest periods, marked with head and spray thirteen times between 1632 and 1850 hours (average interval 10·5 min.) and on another occasion he marked ten times between 0432 and 0650 hours (average interval 13·8 min.). Females were occasionally seen to mark with their heads or necks rubbed into a bush and with scent spraying (Plate 13c), but they never engaged in such a long series of marking actions.

In several instances an animal was seen to react to a recently deposited scent mark by sniffing it and then, sometimes, marking the same spot. On one occasion Scarface sprayed a tree and both Patricia and Romola sniffed the spot immediately afterwards. On another occasion Misty sniffed the bush where Anne had just rubbed her head, then urinated 3 metres away; thirty minutes later she again sniffed the spot Anne had marked, then sprayed it with scent.

Felidae typically cover their faeces with their front feet (Baerends, 1956) but lions were never seen to combine scraping with either fore or hindfeet with defecation, whilst tigers often defecate over a scrape mark (Schaller, 1966).

Tree clawing, described as a group activity earlier in this chapter, may also be a way of marking (Wynn-Edwards, 1962; Ewer, 1968; Leyhausen, 1956a). It is difficult to state whether it indeed has this function in lions, as on only one occasion was any other reaction than imitation observed during this study. In the one exceptional case a one-year old male cub sniffed the spot where an adult female had rested her paws on a tree trunk five minutes earlier.

INTERGROUP RELATIONSHIPS; MEETING OF PRIDES

The relationship of the male to the four prides of females and cubs was discussed previously in relation to the home ranges (Chapter II). He associated with all of them, but was not a permanent member of any one. Whilst strangers are usually not accepted by a pride, the animals will act more antagonistically towards a stranger of their own sex than towards the opposite sex (Schaller, 1969; Adamson, pers. com.; Kühme, 1966).

Among the four prides social contact existed only between Lassie and the Romola pride and Lassie and the Misty pride to which she originally belonged. It is interesting that, after Lassie had split off from her pride, she had a closer relationship with the Romola pride than with her previous companions with whom she was seen only once after her separation.

Lassie was first seen with the Romola pride on 16th August 1968, when they were resting together. Lassie and Chryse were lying close together (1·5 m. dist.) and, 30 m. away, Romola and Patricia were similarly lying in close proximity. When the other three left, Lassie followed them, at a distance, for over one kilometre before separating. Neither Patricia nor Romola seemed to pay any attention to her.

She was next seen with Chryse on 16th September 1968, and in the fifty one days up to 5th November was seen with her four more times. Judging from the cubs' appearance when first seen, Lassie must have been pregnant from about 1st August to 15th November. Outside this fifty-one-day period she was seen only three more times in the company of the Romola pride.

This temporary close association with Chryse was then clearly in connection with her pregnancy. It is interesting though that her companion was not the young female from her own previous pride, Anne, similar in age to Chryse. However, without knowing the past histories of these lions, this is difficult to interpret. Even when she was seen close to Patricia and Romola, Lassie had no amicable social contact with them, only with Chryse, as the Tables on mutual grooming and greeting have shown.

On 5th October 1968, Lassie, Chryse and Scarface were feeding when Patricia and Romola arrived. They were greeted by Chryse but not by Lassie. Patricia and Romola then proceeded to feed, with no show of antagonism from Lassie.

On several occasions the Romola pride was seen feeding on kills that were suspected to have been made by Lassie. There were also some indications that a kill, known to have been made by Lassie, was later shared by the Romola pride.

On 22nd March 1969, Scarface, Romola, Patricia, Chryse and the cubs were feeding in a reed patch which was one of the most frequented places of

Lassie and her cubs during this period. Lassie and her cubs were resting nearby but were not feeding. After Scarface and the Romola pride departed and lay down to rest about 100 m. away, Lassie slowly approached, in a stalking posture, the spot where they had fed, looking towards the pride all the while. She sniffed around the stripped carcase, apparently found nothing edible left, then led her cubs away, passing the pride at 80 m. distance. All looked at her passing by but there was no other reaction. A fresh round wound on Lassie, seemingly made by the horns of an ungulate, suggested that she had either made the kill herself or had participated in it. However it seemed, that while the Romola pride was feeding, she did not dare to approach them. Thus, whilst occasionally sharing a kill, the Romola pride was dominant over Lassie, possibly due to their superior numbers.

Lassie's close relationship to only one member of a pride, in this case, Chryse, is similar to the relationship found in wolf packs, where some outsiders are accepted by part of the pack but not necessarily by all members (Jordan, 1967). This shows the complex pattern of individual relationships that may exist between different groups in both wolves and lions.

On only one occasion, after the split, was Lassie seen close to her former pride, when on 7th March 1969 she led her cubs close to where Misty, Anne and their three cubs were feeding on a young wildebeest. Lassie settled down approximately 100 m. away with her cubs. About an hour later she approached the Misty pride, was greeted by the females, who then left the almost stripped carcase. Lassie picked it up and carried it to her cubs. Thus although Lassie had only infrequent contact with them, the amicable relationship between her and the Misty pride still persisted.

No contact between any of the other prides was ever witnessed.

VI Reproduction and Development

SEXUAL BEHAVIOUR; COURTSHIP AND MATING

It is difficult to determine the time of sexual maturity in a wild population when the exact age of few, if any, of the individuals is known.

Data from captive lions suggest that sexual maturity occurs at about twenty-four months of age for both sexes. Cooper (1942) states that oestrus first appears in lionesses at twenty-four to twenty-eight months. In Lincoln Park Zoo (pers. com.) a pair mated when both were 24 months old. Regents Park Zoo (pers. com.) reported the youngest age for mating for both sexes as thirty-four months.

The mating of the two-year old pair at Lincoln Park Zoo did not result in pregnancy as the youngest age to bear cubs is reported by the same zoo as 30 months and pregnancy is about three-and-a-half months. Youngest age to bear cubs as reported by Regents Park Zoo (pers. com.) was thirty-eight months; for two semi-wild lionesses of George Adamson's (pers. com.) it was thirty-seven and forty-one months.

The oestrous cycle is extremely variable. Cooper reports data on ten females with intervals between oestrus ranging from thirty days to sixteen months with only one out of the ten lionesses showing any regularity in the cycle. Guggisberg (1961) states that the lioness comes into heat every few weeks and Sadleir (1966) reports the mean for nine inter-oestrus periods as 55·4 days.

Although during this study not all matings were necessarily recorded, it is not very likely that many were missed as during these periods the animals are usually in the open and remain in the same area for several days. It is very likely therefore that they will be noticed either by the Park staff or by tourists and reported. The female, for whom most information is available in this respect, was seen mating on the following dates: (1968) 30th, 31st March, 2nd, 3rd, 4th, 17th, 18th September, 22nd, 23rd, 24th, 25th October, 13th, 14th, 15th, 16th December; (1969) 1st, 2nd, 3rd, 4th, 5th, 15th, 17th February. Intervals from the last day of previous mating to the first of subsequent mating range from 10 to 154 days, with no apparent regularity.

Oestrus and readiness to mate may last between seven and fourteen days according to Cooper (op. cit.). Regent's Park Zoo reported this period to be between four and sixteen days when lions are not artificially separated; Lincoln Park Zoo as eight to twelve days and San Diego Zoo as one to three days. Sadleir (op. cit.) gives the range as four to sixteen with an average of 7·2 days.

The longest period for which a pair was seen mating during this study was five days, but the first day or days may have been missed.

The limited data available (fourteen mating periods during sixteen months) in this study do not suggest any preferential season for mating in Nairobi Park. Guggisberg (1961) quotes Ken de P. Beaton, former Warden of this Park, as saying that most lions were born in September and October which would mean a mating peak in May, June and July. Data from European zoos and other parts of Africa suggest that under certain conditions there is a preferred, but not exclusive, mating season. Cooper (op. cit.) found from the data collected at the Gay Lion Farm in California (239 records) that oestrus was more frequent in the spring (February to June).

In lions oestrus is not inhibited during pregnancy or lactation (Schaller, pers. com.). One lioness was seen mating on 6th and 7th December 1968, and in early 1969 was seen with cubs which, judging from their size, must have been born in January 1969. Thus, when last seen mating, this lioness must have been pregnant. George Adamson has made the following observations on two semi-wild lionesses: After the loss of both cubs of her first litter at two months of age a lioness mated again on the day the second cub was lost. After the second litter this same female did not mate again until the cubs were two years old. Another female after the loss of one of two cubs of her first litter abandoned the second cub at about two months and showed signs of coming into season.

Many elements of oestrous behaviour described for captive lions and domestic cats were not seen in the wild lion (Cooper, op. cit.; Michael, 1961). These include increased incidence of head-rubbing and seeking of body contact; pulling up hindlegs with frontpaws and biting them while rolling on back; 'presenting posture' with elevated hindquarters, lowered forequarters and treading.

Schaller (1966) describes a tigress 'apparently in oestrus' spraying scent from her anal glands along her route of travel. Scent squirting was seen occasionally in females but in only five out of sixteen cases recorded could any connection with oestrus be established. In these instances scent spraying occurred (i) during mating, (ii) three or four days before mating, and (iii) sixteen days before mating, thus possibly during oestrus or pre-oestrus.

As most matings occurred well within the range of the female it may be assumed that the male is attracted to the oestrous female rather than the lioness searching for a mate. Scent squirting by the lioness may possibly serve as an attractant. No characteristic 'calling' by females (Cooper, op. cit.) nor rumblings or gurglings by oestrous females, as mentioned by him, were heard, except during actual mating, when a continuous purring sound was emitted.

A mating pair of lions may or may not separate from the pride (Guggisberg, op. cit.). During this study mating pairs were most often found alone, but, on several occasions, they joined or were joined by the female's pride sometime during the mating period and then continued mating while other members of the pride surrounded them. (Plate 18 c). Kühme (1966) saw mating lions in close proximity to a pride including other adult males. There seems to be a tendency in young females to try to rejoin their pride during mating. This was noticed with two of the youngest females on three occasions. The females gradually moved closer and closer to their prides until they joined them. They walked short distances (less than 100 m.) between copulations, always closely followed by the male. Generally the mating pair stayed within an area of approximately 1000 m². during mating. The short walks between copulations were a frequent occurrence, but in most cases did not result in any real displacement as the pair tended to move in a circle with the animals ending up, after several hours, in their original position.

The typical sequence of events during a period of mating is as follows. The two animals are resting close together (Plate 14 a). Either one may initiate the sequence leading to copulation. The male often starts the sequence by a 'mating snarl' (Guggisberg, op. cit.) described by Cooper (op. cit.) as a 'sneeze-like grimace'. This display consists of the animal pulling back his lips, baring his teeth, wrinkling his nose and rolling his head slowly from side to side without a sound, while facing in the direction of the female. This snarl-like expression (Plate 14 b & c) is indistinguishable from a snarl in agonistic behaviour except for the slow rhythmical performance of the behaviour pattern. The mouth is opened slowly, not suddenly, as in most cases in an agonistic snarl and the movement from side to side is always slow and rhythmical. This change in speed and the reiteration of the movement is a characteristic feature of ritualization (Ewer, 1968). Thus this mating snarl may be regarded as a ritualized display with a signal function, as the female almost invariably responds by getting into the crouching position for copulation.

If the female fails to respond, because she did not look in the direction of the male during this display, the male may get up, approach her and give a few strokes of the tongue usually in the shoulder, neck or back region (Plate 15 a). These few strokes of the tongue are a symbolic form of social grooming for there are never more than one or two licks. They seem to have a signal function similar to the mating snarl, for the female responds in the same way.

The lioness may react to either display by just getting into the crouching position for copulation or, more frequently, by walking a short distance before crouching down. The male, growling continuously, follows her closely, (Plate 15 c) and mounts immediately after she crouches. Occasionally he may put a paw onto her back as if to push her down into the crouch (Plate 15 b).

This close following, whenever the female moves is a very characteristic behaviour of the male during courtship and mating (Guggisberg, op. cit.) and is often the first sign that a lioness is coming into oestrus. As the mating period draws to its close, however, the distance at which the male follows gradually increases until, by the last day of mating, it may be as much as 25–30 m. Failure to follow the female persistently indicates the end of the mating period.

Another characteristic behaviour pattern connected with this close following of the lioness is the male's tendency to interpose himself between his mating partner or oestrous lioness and any other animal. This was observed on many occasions when Scarface stood between his mating partner, or a female in oestrus, and another female or cub trying to approach the lioness, and showed agonistic behaviour. Similar behaviour was observed in wolves where the male interposes himself between his female and other males, but no mention is made of his reaction to other females (Rabb et al., 1962).

As soon as the female crouches, the male mounts and executes rapid thrusting movements. Towards the end of this sequence of thrusts he opens his mouth in a wide biting movement taking the female's head or nape between his jaws, with the teeth touching the skin but not pressing down (Plate 16 a–c). This may be done once or several times on alternate sides. The female does not usually react to this 'neckbite'.

After the thrusts have stopped, and presumably during ejaculation, the male emits a long-drawn-out howling roar, then jumps off or, more rarely, slowly disengages himself from the female. On a few occasions a lioness was seen to snap or lash out towards the male while he was dismounting. However, in the majority of observations, there was no aggressive display by the females, in contrast to observations on domestic cats and captive lions where the females show a high level of aggressive behaviour (Michael, 1960; Cooper, (op. cit.).

In cases in which the female was the first to move from the resting position, she usually circled the male several times (Plate 17 a–d) rubbing her body against his and occasionally lifting her tail sideways before crouching down. Sometimes she simply walked away a few steps and then crouched down into the copulatory position.

After copulation the lioness rolls onto her back, or walks away a few steps before rolling, then usually remains in a lying position, either stretched out on her side or on her haunches. While she rolls, the male stands close to her and only lies down close to the female after she has stopped rolling (Plate 18 a–b).

From the time when she crouches down (or shortly before) until the roar of the male makes it impossible to distinguish, the lioness emits a steady sound like a loud purring or gurgling. Cooper (op. cit.) mentions a 'gurgling-rumbling' sound, made by the female, which increases in volume. Guggisberg (1961) states that he never heard any sound other than an occasional grunt, growl or snarl from mating lions. As this study is based on the same male mating with different females, it is difficult to establish which behaviour elements are constant and which exhibit a greater individual variation. However, purring was heard from all seven females that were seen mating.

Cooper's (op. cit.) description of copulation in captive lions differs in many respects from that described above. The main reason for this may be the separation of the sexes in captivity, and their being allowed together only for a short period for mating. Under these conditions females seem to behave in a much more aggressive manner, both before and after copulation, than the lionesses in this study. The 'presenting position' with elevated hind-quarters believed by Cooper to be a releasing stimulus for mounting was never observed and the only requirement for the male to mount was for the female to assume the crouching position.

The violent rolling as an after-reaction to copulation observed in the female domestic cat (Michael, op. cit.) was also part of the lioness' post-copulatory behaviour. However, no lioness was ever seen to exhibit male patterns of behaviour, such as neckgrip, mounting and thrusts, as seen in captive lionesses (Cooper, op. cit.). This behaviour may also be an artifact due to the separation of the sexes in captivity.

Table VI–1 lists the elements of mating behaviour described above and the frequency of their occurrence.

Table VI–1
Elements of mating behaviour and their frequency of occurrence in 146 matings

Behaviour elements	Observations	
	No.	Per cent of total
1. Male initiates	53	36·3
2. Female initiates	84	57·5
3. Male snarls	35	23·9
4. Male follows female	79	54·0
5. Female circles male	12	8·2
6. Female rubs against male	13	8·9
7. Male licks female	64	43·8
8. Tail-flick, male and female	18	12·3
9. Female purrs	66	45·2
10. Male paw on female	11	7·5
11. Neckbite once	69	47·2
12. Neckbite more than once	51	34·9
13. Male stands after copulation	94	64·3
14. Female rolls on back after copulation	82	56·1
15. Male roars	133	91·0

The whole sequence from the time the first animal gets to its feet until the second animal resumes the resting position usually lasts between 30 and 70 seconds. In 17 observations the range was 30–154 seconds with 55 seconds as the average. It seems that any movement by the female during the rest period acts as a stimulus on the male to start the sequence leading to copulation. Sitting up, yawning, scratching, stretching, grooming, may all act as triggering actions.

The frequent repetition of the mating act is characteristic of the lion (Guggisberg, 1961). Out of a total of 191 daytime observations on intervals between matings, the majority, 54%, fell between 8 and 18 min.; Kühme (1966) found the average interval to be 17 min. For forty-four night observations 56% fell between 7 and 25 min. Intervals ranged from 7 to 148 min. by day and from 7 to 131 min. by night. Length of interval was calculated from the time the second animal lay down to rest to the time the first animal got to its feet to initiate a new mating sequence.

An interesting feature was the almost identical long rest period during two night observations, between 0324 hours and 0535 hours and between 0350 hours and 0541 hours, 111 and 131 min. respectively. The next longest interval for night observations was 42 min. With the limited number of night data available it is not possible to establish the significance of this phenomenon. During the day the longest periods of rest tended to occur around midday when the animals seek shade and reduce the level of all activities.

A varying degree of aggression is found in most animals' mating behaviour as an element necessary for breaking down the individual distance barrier between the mating pair (Walther, 1963). At the more primitive levels of mating behaviour this aggression is more overt, while at a more highly evolved stage aggression tends to be manifested only by symbolic actions.

Lions are contact animals, hence the breaking down of the individual distance barrier does not necessitate much aggressive behaviour. In fact, very little aggression was observed during mating. The mating snarl is ritualized and the neckbite is a symbolic action as the male does not actually press with his teeth. Thus the aggressive elements in their present form indicate that mating behaviour in lions is of a highly evolved type.

Aggression by the female was observed only infrequently and on these occasions the male merely evaded, without showing any aggressive response himself.

During most of the mating periods the pair remained together for several days without feeding. The food seeking drive is inhibited during this time by the sexual drive (Baerends, 1956). Towards the end of the period, however, the food seeking drive gradually reasserts itself. This is noticeable especially in the female who at this time evinces a gradually increasing interest in prey animals that appear within sight of the pair. On 14th January 1969, a mating pair seen together for the third day running was resting 30 m. apart. This distance already indicated that the male's following drive was lessening and thus the mating period was approaching its end. They were seen mating at 1025 hours for the last time, when the female snarled and snapped towards the male as he jumped off after copulation. The female walked away, stopped to drink, then walked on and at 1055 hours successfully stalked a young warthog. She then carried the warthog away and disappeared with it in a thicket. The male followed her at about 100 m. distance but, when she disappeared, turned and walked back in the direction from which he had come. This behaviour of the male was in striking contrast to his behaviour on other occasions when a lioness had made a kill in his presence. On all other occasions the male had appropriated the carcase immediately. Possibly the food seeking drive of the male on this occasion was still inhibited by his sexual drive, while that of the female had already displaced her sexual drive.

On another occasion, however, when a young kongoni was killed by two females in the company of a mating pair, the male appropriated the kill in his usual manner, proceeded to feed for 80 min. and resumed his mating activity as soon as he had finished feeding.

Often after copulation the male flicked up his tail, either once or several times in succession, with considerable force (Plate 8 b). This behaviour was also observed under other circumstances (see Chapter IV). The female was occasionally seen to flick her tail up in a similar manner but she more often lifted her tail up and sideways before copulation. This was similar to the tail lifting described as part of 'presenting activity' by Cooper (op. cit.) but did not appear to be a stimulus for mounting as claimed by that author.

GESTATION PERIOD, LITTER SIZE, CUB MORTALITY

According to observations made by the London Zoo as quoted by Guggisberg (op. cit.) actual fertilization of the egg cells occurs only on the fourth day of oestrus. Although it is not certain whether this applies to lions in the wild, it appears that the majority of matings do not result in pregnancy (Schaller, pers. com.). Of fourteen mating periods observed during this study only 4 or 28% seem to have resulted in fertilization. Cooper reports (op. cit.) ninety-three fertilizations out of 239 breedings, or 38% successful matings.

Zuckerman (1953) quotes the following gestation periods from the literature: 100 to 113 days (Sclater, 1900), 108 days (Blandford, 1891), and 111 days (Brown, 1936). A period of 105 to 110 days was reported from the Lincoln Park Zoo (pers. com.) and 108 to 120 days from Regents Park Zoo (pers. com.). Cooper gives the average for fifty-one pregnancies as 109·74 days. He also notes that, in a few cases, the first gestation periods were longer than subsequent ones in the same lioness.

As no cubs younger than about two months old were seen during this study neither length of gestation period nor actual size of litters at birth could be established. Litter sizes as reported from zoos and as seen, when two months old, during this study and sex ratios of litters are shown on Table VI–2.

Table VI–2
Litter size and sex ratio

| Total No. of litters | Litter size | | Source | Sex ratio | |
	No. of young in litters	Average No. of young		Per cent male	Per cent female
23	1–4	2·35	Bronx Zoo, N. Y. pers. com.	46	54
67	1–6	2·64	Regents Park Zoo pers. com.	48	52
17	1–3	2·70	Lincoln Park Zoo pers. com.	60	40
5	1–4	2·20	San Diago Zoo pers. com.	36	64
3	2	2·00	G. Adamson pers. com.	50	50
?	1–5	2·5	S. Zuckerman, 1953	?	?
64	1–5	2·48	Cooper, 1942	?	?
6	3–4	3·3	This study	50	50

All zoo data combined

| | Number of cubs in litter | | | | | | Total cubs |
	1	2	3	4	5	6	
Number of litters	15	38	44	11	3	1	288

Most frequent number of cubs in litter: 3
Average number of cubs in litter: 2·6
Total number of cubs sexed: 234

Per cent male	Per cent female
47·5	52·5

Of the six litters seen at two months old, four had three cubs and two had four cubs. However, six months later one cub was lost from each of the four-cub litters and one from a three-cub litter.

According to the literature and information from zoos the size of litters varies between one and six, three cubs being the most frequent. An exceptional case of seven cubs was reported from Arnhem Zoo, Holland (van Hooff, 1965). Table VI–2 shows average sizes in zoos ranging from 2 to 2·7. The average for Nairobi Park was 3.3. Mitchell et al. (1965) reported a steady decline in Kafue Park, Zambia, from an average litter size of 4·5 in 1958 to 2·00 in 1963. He believes this may be the result of an influx of lions into the Park from the surrounding areas. Stevenson-Hamilton's (1937) contention that litter size in lions increased from two or three to four or five with increasing food supply in Kruger Park would lead one to expect larger litters from well-nourished animals in zoos.

Cub mortality is quite high in captive lions, especially in first litters. Mortality during the first six months in captive animals ranges from 9% to 72% (Table VI–3). For the reasons stated above, total post-natal mortality could not be determined in this study, but the loss between about two and six months was three out of twenty, or 15%. None of the lost cubs seem to have belonged to a first litter as far as could be judged from the apparent age of the mother.

The higher cub mortality in first litters as reported by Regents Park Zoo (pers. com.) and George Adamson (pers. com.) may be partly due to the mother's lack of experience. Young cubs may be taken by leopards or adult male lions or may be abandoned by the mother for no apparent reason (Adamson, pers. com.). Adamson tells of two semi-wild lionesses which both abandoned the second cub in their first litters after one cub was lost at two months of age. In cases where a litter is lost or abandoned the lioness may immediately come into season again and thus produce another litter within a year. In captivity, where cubs are usually taken away, lionesses may produce litters at five- to six-month intervals. In Regents Park Zoo a lioness which lost her cubs at thirty-two days produced a litter 130 days after the first litter was taken from her.

Cooper (op. cit.) also reports higher mortality in first litters when the causes may be the refusal of the lioness to care for her young and occasional cannibalism by the mother. In wild populations cannibalism has been reported mostly on the part of male lions (Schaller, 1969; Adamson, pers. com.; Cullen, 1969; Guggisberg, 1961). Some losses also occur through other predators or large herbivores; hyaena and leopard occasionally kill cubs; buffalo and elephant have been known to trample them to death (Makacha and Schaller, 1969; Cullen, 1969; Schaller, 1969). However, Guggisberg (1961) states that mortality in young cubs is not very high in good game country and cites the Mara Reserve and the Athi Plains, including the Nairobi Park area, as examples where litters of up to five cubs were successfully

Table VI–3
Cub mortality up to six months of age

No. of litters	No. of cubs in all litters	Mortality		Source	Reference
		No.	Per cent		
4	?	—	9	New York Zoo	pers. com.
67	177	58*	32·77	Regents Park Zoo	pers. com.
17	46	6	13·04	Lincoln Park Zoo	pers. com.
5	11	8	72·73	San Diego Zoo	pers. com.
3	6	3	50·00	G. Adamson	pers. com.
6	20	**3	15·00	This study	—

* Most in first litters.
** One of these cubs was apparently abandoned and adopted by the Park Warden.

raised. He believes that the greatest losses among juvenile lions in Nairobi Park occur when they become independent and start roaming outside the Park where they may be killed by hunters, poachers or farmers.

When conditions are less than optimal, litters become smaller, cubs may fall victim to increased aggression during feeding time, cannibalism may increase, starvation may decimate the cubs (Guggisberg, op. cit.; Schaller, op. cit.). Both Guggisberg and Schaller agree that only about 50% of young survive to adulthood, even under good conditions.

This high mortality of juveniles is a mechanism of population control, and is similar to the high pup mortality in wolves, another predator with few, if any, natural enemies, except man (Jordan, 1967).

PHYSICAL DEVELOPMENT AND CARE OF YOUNG

Most herbivores are born in an advanced stage of development and very shortly after birth, in some cases within a few minutes, are able to walk and keep up with the herd. In contrast, predators are usually born helpless and need an extended period of parental care. Lions are born blind and with limited locomotory powers (G. Adamson, pers. com.) weighing only 100th of the mother's weight, or about 1·5 kg. (Average of twenty-one cubs 1,400 gr.; range 1,134 gr. to 1,700 gr. Lincoln Park Zoo, Chicago, pers. com.). By two months they weigh about 4 kg. and are able to run about. (Average of ten cubs 4,180 gr., range 3,150 gr. to 6,550 gr. Lincoln Park Zoo, Chicago, pers. com.). Plate 20 c shows a cub of about two months of age alongside its mother.

The lioness leaves the pride for the period of parturition and gives birth in a secluded spot, a rock shelter or a thicket. For the first six to eight weeks the cubs stay in the shelter and the mother divides her time between her pride and the cubs. Usually at about two months of age the cubs join the pride (G. Schenkel, 1966; G. Adamson, pers. com.). But this seems to hold true only when other cubs already present in the pride are no more than about four months old. In one case where the cubs in the pride were over four months old, the mother (Lassie) broke away from the pride; in a second case, where the cubs in the pride were thirteen and fifteen months old, the mother (Chryse) only introduced her cubs to the pride when they were four months old. When cubs are no more than two months apart in age they are usually all brought up together from the time the younger ones reach the age of two months, this was the case with the cubs of Patricia and Romola.

Young lions, as a rule, open their eyes between the fourth and sixteenth day (Cooper, 1942; David, 1962; Forbes, 1963). Occasionally, when the gestation period is very long, they may be born with their eyes open, but, until the fourth week, vision appears to be limited to light perception (Forbes, op. cit.; Cooper, op. cit.).

During the first week cubs can only drag themselves along by their forefeet, as co-ordination of the hindquarters does not develop until the second week to the stage where the cubs can start walking unsteadily. By the fifth week they can walk well and can trot, and by the eighth week they are able to run (Cooper, op. cit.; Forbes, op. cit.).

The first teeth appear between the fourteenth and twenty-second day and by about two months dentition is complete, including molars (David, 1962; Mrs Bill Woodley, pers. com.).

According to data on twenty-three cubs from Lincoln Park Zoo (pers. com.) there is no difference is size between male and female cubs up to about six months but between six and twelve months the males begin to broaden in shoulder and paws and by one year old they are noticeably bigger than their litter sisters (Cooper, op. cit.). This is also true for the Nairobi population.

FEEDING AND GROOMING OF CUBS

Lion cubs suckle for an exceptionally long period; during this study cubs were seen suckling occasionally up to fifteen months of age. Tigers, in contrast, are weaned at about five to six months (Schaller, 1967). There is no definite weaning period when the mother discourages suckling by a show of aggression as is the case in some apes (Goodall, 1967) but the frequency of suckling gradually decreases until it fades away completely, probably due to the mother's milk drying up. Thus it is probable that cubs in a pride with a younger litter present may be able to suckle for a longer period than if they were alone, as a lioness will suckle any young cub in the pride. This leads to competition between the larger and smaller cubs and probably acts, under unfavourable conditions, as a limiting factor on the population. Patricia's smaller cubs were often seen to be pushed away by Romola's larger cubs when they tried to suckle from either lioness. It was not always possible to determine which of the cubs was suckling, but the limited number of observations (total of sixty-three) suggests that Romola's larger cubs suckled from Patricia more than half as many times as did Patricia's own cubs, while the smaller cubs suckled Romola less than half as many times as did her own cubs. All cubs also tried occasionally to suckle from Chryse, who had no young at that time, but were always discouraged by the young lioness who either pushed them away or made a slight display of aggression, such as snarling.

Length of suckling by any one cub ranged up to twelve minutes with an average of eight minutes. Of fifty observations on suckling only one fell between 1400 hours and 1600 hours; the rest were equally divided between 0800 hours to 1400 hours and 1600 hours to 1900 hours.

Although during the day cubs suckle frequently when their mother is with them, they are able to go for long periods without milk, as the mother is often away hunting for the whole night and may even leave them alone for up to twenty-four hours (Schaller, 1969). Lion's milk is very concentrated with a high fat content and high satiety value; thus nursing need not be frequent (Ben Shaul, 1962).

In kittens and pigs definite teat ownership exists and the releasing stimuli are believed to be olfactory (Ewer, 1959). In lion cubs which suckle from any lioness in milk, such teat ownership must, of course, be absent. However, olfactory stimuli in a negative sense, in the form of a repelling mechanism, may be present. This may be similar to a suspected repelling mechanism in anoestrous female domestic cats (Michael, 1961). This assumption is based on three observations when a cub approached a teat as if to suckle, then, after bending its head close to it but without touching, turned away without feeding. Similar behaviour was observed in a young chimpanzee, but its reaction was more intense, as it turned away screaming (Goodall, 1967). In neither case was any threat from the mother observed. One would assume such a repelling mechanism, if it exists, to be functional towards the end of the lactating period; however, these observations were made on 8th, 22nd October and 11th November 1968 when the cubs of the two lionesses were six and eight months old. As previously mentioned cubs suckled with decreasing frequency, for over a year. However, this six–eight months period approximately corresponds to the time when tiger cubs are weaned.

Cubs are usually led to a kill; only once was a lioness seen to carry a carcase to her young. The size of prey may affect the course of action the mother takes. Cubs are sometimes taken long distances to a kill. Lassie's cubs at ten weeks walked 3 km. and at fourteen weeks were led for 1½ km. with only four resting periods lasting no more than two minutes each.

A mother does not fetch her young cubs immediately general feeding starts, but only after the pride has eaten for some time. This presumably serves to shield small cubs from injury during the initial stages of feeding when competition for food is at its highest intensity. Thus Patricia collected her cubs only after the adults and Romola's older cubs had been on a wildebeest kill for several hours.

Carr (1962) maintains that the lioness regurgitates meat for her young, but neither Adamson (pers. com.), nor Schaller (1969) believe that she does. Regurgitation was never observed during this study.

Besides suckling, the most frequent interaction between a lioness and her cubs is grooming. In social grooming only certain parts of the partner's body are groomed (see Chapter V), however, the mother grooms the whole body of her young (Walther, 1963). As in social grooming, utilitarian and social functions are combined. Licking has a general cleaning and stimulating effect, and in the first five weeks it also serves as a necessary stimulus for defecating and urinating (Forbes, 1963).

All three females in the Romola pride groomed all cubs, but the mothers groomed their own offspring more often than those of another lioness. Table VI-4 shows the relative percentages of grooming by the three lionesses. As one litter had three and the other two cubs, percentages were calculated per cub to make the figures comparable. Both Romola and Patricia groomed their own cubs more often, but the difference is bigger in Patricia, where the ratio is about 4:1, while for Romola it is less than 2:1. Chryse groomed the smaller cubs more than twice as often as the larger ones; possibly smaller cubs offer a stronger stimulus to the maternal instinct of a female who is not the mother, than larger cubs do.

Table VI-4
Adult females grooming cubs in Romola pride

Active / Passive	Romola			Patricia			Chryse				Total	
	No.	Total	Per cent per cub	No.	Total Per cent	Per cent per cub	No.	Total Per cent	Per cent per cub	No.	Per cent	Per cent per cub
Romola's 3 large cubs	54	69.0	23.0	15	26.4	8.8	11	39.6	13.2	80	49.2	16.4
Patricia's 2 small cubs	24	31.0	15.5	42	73.6	36.8	17	60.4	30.2	83	50.8	25.4
Total	78	100.0	23.0×3 / 15.5×2	57	100.0	8.8×3 / 36.8×2	28	100.0	13.2×3 / 30.2×2	163	100.0	16.4×3 / 25.4×2
			100.0			100.0			100.0			100.0

PROTECTION OF CUBS

Protective behaviour in a lioness manifests itself mostly in shielding her young from other lions and in preventing them from straying too far from their companions. In many species of mammals adult males are kept away from the young. Young chimpanzees are prevented from contact with all adults but especially adult males and often even the older offspring of the mother (Goodall, 1967). The lioness also keeps her young away from adult males as well as adult females of other prides but not from females of her own pride. Cubs under four months also seem to be kept away from much older cubs.

When one of her young is the subject of aggressive threat the mother will call it away. This was observed on several occasions when Scarface snarled and growled at a cub and the mother approached the cub, called it and then led it away. In no case did a mother respond to a cub's distress call but only to the aggressive display of the male. It is interesting that there appears to be no reaction of a mother to the distress call of a cub. This was noticed when a two month old cub of Patricia's, found abandoned, was returned to the pride and put down in front of its mother, Romola, and the other cubs. It miaowed and whined loudly but neither lioness reacted. The other cubs, however, approached, nuzzled and greeted it.

This lack of reaction by the mother to a young's distress call was also noted by Ewer (1968) in Suricata where only littermates respond in the way these lion cubs did. She interprets this as serving to keep the litter together in the mother's absence. It may well have the same function in lions, where a cub is usually either within sight of the mother, or out of hearing as well as sight. Thus reaction by the mother to a cub's vocal distress is not important, while it is important to keep the litter together when the cubs are left on their own. (Plate 20 b).

Often when cubs were disturbed, as when a vehicle approached too closely, they ran to their mother and huddled close to her. This they did even when they were sitting close to another adult when the disturbance occurred. Occasionally they reacted to a disturbance or threat display by Scarface by running to the mother and suckling for a very short time. Lassie's cubs at six months and Bertha's at five months were observed to behave in this manner. This may be similar to the 'reassurance nipple contact' observed in frightened young chimpanzees by Goodall (op. cit.).

VOCAL COMMUNICATION BETWEEN MOTHER AND CUB

Whilst the mother does not react to a cub's distress call, the cubs up to eight to 10 months of age almost invariably respond to the mother's call. This cub-calling is a specific hollow, short sound, repeated several times at low or medium intensity, (Table V–5, 16). The cub-calling sound is used (i) when cubs wander too far away, (ii) to call cubs away from conspecifics (iii) when the mother starts walking and wants the cubs to follow, and (iv) when the mother returns to the cubs. In the latter case it is usually heard by the cubs a few minutes before the lioness appears. In the first three cases the cub's reaction is to run to the lioness immediately, in the fourth case cubs under about four months tend not to move towards the sound but to sit up, look intently in the direction of the sound and only run towards the lioness when she is in sight. When they are older, they may start running towards the sound before seeing her. This cub-calling is not specific, as Romola's cubs responded to Patricia's call and vice versa. Chryse never uttered this particular sound before she had cubs of her own.

Cubs may fail to respond to the sound if they are in the company of another adult lion when the call is heard. Thus Lassie's cubs at six months old were sitting near Scarface and did not react to their mother's call when a sudden noise caused her to call them. However, a short time later, when a vehicle approached them too closely, they all ran towards the lioness and huddled close to her.

Whether cubs follow the departing mother, or stay behind, is probably only determined by her calling them or failing to do so. On 22nd March 1969, Lassie's cubs (four months old) were playing when the mother walked away. The cubs stopped playing, took a few steps in the direction the lioness was going, then sat down and looked after her. Similar cases, when cubs did not follow unless the cub-calling sound was given by the lioness before moving off, were frequently observed.

PLAY AND THE DEVELOPMENT OF BEHAVIOUR
PATTERNS AND SOCIAL INTERACTIONS

Many young mammals play but the felids are especially noted for the frequency with which their young engage in this activity. Elements from various behaviour patterns occur in play but they lack the ordered sequences of the original. Moreover elements from different functional contexts such as prey catching, escape or fighting may be interwoven. Other characteristics of play are the exuberance of movement and the lack of consummatory stimuli acting as feedback and bringing the course of action to an end (Ewer, 1968).

Some of the functions of play are: physical exercise, an opportunity to experiment with innate action patterns and gain experience in the various ways in which they can be used and general enrichment of experience (Ewer, op. cit.; Leyhausen, 1965 b). Social play also serves to strengthen the bond between mother and young and between siblings (Goodall, 1967). This latter aspect is probably especially important in felids in order to keep the littermates together during the mother's protracted absences (Ewer, op. cit.).

The type of play that is most prevalent depends on the species and its way of life. In non-predators escape behaviour is often the most important play element; such is the case in the ground squirrel, but in predatory species such as the felids, fighting and prey catching elements are predominant (Ewer, op. cit.).

Play, common amongst cubs, declines in frequency after the first year. It occurs occasionally between adult females, but is almost non-existent in adult males (Cooper, 1942; see also Chapter II).

Social play was more prevalent than solitary play in both adult females and cubs. Normally the play of the adult females involved the cubs but sometimes they played with each other, chasing and jumping onto the partner and biting playfully. These instances of lionesses playing with each other usually occurred between 1800 hours and 0800 hours. The youngest females played most frequently, even playing with the cubs more often than their mothers did. In Romola's pride Chryse was seen playing with the cubs on sixteen occasions while the two mothers were seen only six times each. In Misty's pride, Anne played five times with the cubs while their mother was only once seen to do so.

When a lioness plays with cubs she will usually paw or bite gently and occasionally butt with her head. Lionesses also occasionally swish their tails up and down apparently inviting the cubs to play: the cubs then try to catch the tail by pouncing upon it. This catching of a moving object may be the first manifestation in play of the prey-catching pattern.

In the cub's play amongst themselves no preference could be established between littermates as against the two litters in Romola's pride.

Elements from the food getting pattern of behaviour and of aggressive behaviour were the most common, such as stalking, jumping onto the partner, biting and 'boxing' (Plate 19 a, b) when cubs hit each other with their paws (element vi of threat intimidation, Chapter V). This hitting with the paw, listed by Cooper (op. cit.) as the most important element in fighting, was only rarely seen in its original context in this study, but occurred frequently in play.

Biting was directed most often towards the back of the partner's neck, the typical 'killing bite' of some Felinae (Leyhausen, 1956 a). In this connection it is interesting to note that none of the prey killing techniques most frequently observed—hold on throat or muzzle—(see Chapter VII) was ever seen during play. The bite in the nape of the victim's neck seems to be an element of intraspecific fighting rather than of prey-catching, as most leopards and lions found killed by other lions were bitten through the back of the neck (Guggisberg, 1961; Cullen, 1969).

Ewer (op. cit.) speculates as to whether a newly evolved behaviour pattern appears in play as soon as it has evolved or, if not, how long it takes before it becomes incorporated into the species' repertoire of play activities. The killing muzzle hold of the lions, described more fully in Chapter VII, may be one such pattern that has not yet appeared in play.

As Ewer (op. cit.) notes, it is not always easy to distinguish between play and the first immature manifestation of an innate action pattern. Thus, the stalking, first observed in a five-month-old cub, and the use of grass as cover for stalking by a six-month-old cub were probably examples of play, as the object of the stalk, in each case, was another cub. However, when eight-month-old cubs first stalked other species—in one case an eagle and in the other a zebra—we probably have attempts at actual prey catching. Two fifteen-month-old female cubs, seen to stalk co-operatively and to try to encircle their prey, were displaying the first evidence of social hunting behaviour. The exuberance of movement in the first two cases, and the absence of this characteristic in the latter cases confirm this impression.

As Guggisberg noted (1961) female cubs are more active in hunting attempts; no attempts by the male cubs at co-operative stalking were ever seen and when two twelve-month-old females took part in a stalk by the three adult lionesses, their litter brother and the two younger cubs (ten-month old) stayed behind.

Solitary play was less frequent than social play. It was seen as early as the second month when cubs played with loose pieces of paper, pebbles, a piece of rubber, a stick or, in fact, any loose object they could find (Plate 19 c). They often played with a branch (or a bunch of high grass) by pulling it down repeatedly and letting it go again.

They were also occasionally seen playing with a carcase before starting to feed, especially before the soft inner parts were exposed, as they could not break through the skin unaided. In one instance five and seven-month-old cubs played for over an hour with a zebra foal's carcase before a lioness carried it into a nearby patch of vegetation where it was subsequently consumed. The young adult, Chryse, was also occasionally seen playing with a carcase before feeding. Play with dead prey has also been described in cats (Leyhausen, 1956 a). Ewer states that this is likely to occur after a difficult battle, when the predator's general excitement has been raised to a high level of intensity and cannot be 'switched off' immediately the prey has died. However, in lions only young animals—never lionesses which did the killing—were observed to play with dead prey.

Sexual behaviour patterns were occasionally seen: mounting was first seen from a five-month-old female cub, and mounting with accompanying thrusts was first seen from a six-month-old male cub. No females, neither adults nor cubs in play, were ever seen to perform copulatory thrusts; this is in marked contrast to the apparent frequency of such behaviour in captivity (Cooper, op. cit.).

For the first few months of their life in the pride the cubs in Romola's pride had occasional contact with the adult male, but showed defensive behaviour or flight at his approach. They gradually became less shy and by the fifth month occasionally rested close by him with no sign of fear. However, any attempted bodily contact, such as headrubbing, by the cubs was actively discouraged by Scarface. The cubs in Misty's pride, with which Scarface had less frequent contact than with the Romola pride, even showed flight reactions to the male at eight months of age.

The earliest social behaviour pattern evident in cubs was headrubbing. This was seen when the cubs were first observed at about two months, when it was often performed towards the mother as contact seeking or reassurance contact whenever any disturbance occurred. The mother usually responded by grooming the cub for a short time. Cubs greeted other lionesses in the same way, from about the same age. Table VI–5 shows that a cub greeted its own mother most frequently and the young lioness, Chryse, the least often. Romola's cubs first greeted Scarface at nine months of age, but the younger cubs were never seen to do so even by the end of this study when they were thirteen months old.

Greeting towards another cub was not noted until the fifth month. The older male cub, Calef, was greeted most often and on the average the older

**Table VI–5
Romola Pride:
cubs' headrubbing
to adults**

Active ⟍ Passive	Romola's 3 large cubs No.	Per cent per cub	Patricia's 2 small cubs No.	Per cent per cub	Total No.	Per cent per cub
Scarface	4	1·0	—	—	4	3·0
Romola	42	10·3	12	4·5	54	39·9
Patricia	36	8·9	25	9·3	61	45·3
Chryse	9	2·2	7	2·6	16	11·8
Total	91	22·4	44	16·4	135	100·0
Total Per cent	3 × 22·4 67·2		2 × 16·4 32·8			100·0

cubs were greeted more often than the younger ones (Table VI–6). But the figures are small and so no definite conclusions should be based on them.

Although self-grooming occurred as early as the second month social grooming of lionesses appeared later, at three months, while another cub was groomed for the first time at five months. Cubs did groom lionesses other than their mother but less frequently. Table VI–7 shows grooming percentages calculated on an individual cub basis. Mothers were groomed about twice as often as the other lioness with cubs, but Chryse was groomed much less frequently. Although her cub-grooming performance was about half as much as Patricia's (see Table VI–4) she received less than one quarter the attention Patricia did.

The larger cubs groomed the lionesses more often than the smaller ones, although Table VI–4 shows that they got less grooming attention than the younger cubs.

Aggressive displays between cubs were observed when the cubs were first seen. However the first sign of aggression against an adult was not seen until the eighth month in the form of a snarl towards the mother. But such displays remained rare up to the end of the study period, when the largest cubs were fifteen months old, and were still limited to snarling and snapping, without growling (see Chapter V threat display iii and iv).

Up to the end of this study no cub was heard roaring, neither was spraying of scent with uplifted tail nor scraping while urinating ever observed. However, marking with the head was occasionally seen from the ninth month.

DEVELOPMENT OF SEXUAL DIMORPHISM AND INDEPENDENCE

From about the ninth month a male cub begins to show signs of behavioural as well as physical dimorphism. At this time the mane begins to grow, first as a slight ruff around the neck then as a crest along the top of the head and the back of the neck. In the pattern of self-grooming the growing mane on the chest progressively receives more and more attention compared with any

**Table VI–6
Headrubbing
(greeting) cubs to
cubs**

Active ⟍ Passive	Calef m.	Karen f.	Carla f.	Victor m.	Pamela f.	Total
Calef	—	1	2	5	5	13
Karen	3	—	2	2	2	9
Carla	—	1	—	1	—	2
Victor	1	1	—	—	1	3
Pamela	—	—	—	1	—	1
Total	4	3	4	9	8	28

Table VI–7
Cubs grooming
adult females in
Romola Pride

Active Passive	Romola's 3 large cubs		Patricia's 2 small cubs		Total	
	No.	Per cent per cub	No.	Per cent per cub	No.	Per cent per cub
Romola	29	14·2	5	3·7	34	10·0
Patricia	18	8·8	11	8·1	29	8·5
Chryse	4	2·0	1	0·7	5	1·5
Total	51	25·0 (3x)	17	12·5 (2x)	68	20·0 (5x)
Per cent		75·0		25·0		100·0

other part of the body, the emerging grooming pattern of the adult male. Male cubs are usually more alert and are often the only members of the pride to sit while the others lie just as Scarface was shown to be more alert than the lionesses (Table III–1). At the same time male cubs tend to keep themselves apart from the other members of the pride and to rest at some distance from the group. They also become the last to start moving and stay at the rear of the line. These tendencies, which were noticed from the ninth month onwards, increased in degree until observations ceased.

It is possible that the higher mortality of juvenile males resulting in a preponderance of females (see Chapter II) as found in most areas, may be partially the consequence of a premature development of this pattern. One of Patricia's male cubs at two months of age repeatedly strayed away from his littermates and was finally adopted by the Game Warden. This may have been an example of such premature independence.

In monkeys male infants show earlier signs of independence than female young, but this originates in the mother's differential behaviour towards her male infant, as she keeps the females more close to herself than the males (Jensen, et al., 1968).

In chimpanzees an adolescent male gradually becomes dominant to other females while still subordinate to his mother (Goodall, 1967). In a similar way a male cub gradually shows less amicable behaviour towards adult females other than his mother. Thus, on 28th April 1969, when Romola approached the five cubs, they all walked to her and rubbed their heads to her in greeting. When, an hour later, Patricia approached them, all cubs except Calef greeted her. Thus Victor, an eleven-month-old male, still greeted an adult lioness not his mother, while Calef, at thirteen months did not do so any more.

Tiger cubs show signs of independence shortly after being weaned at about six months of age (Schaller, 1966), by eleven months begin hunting on their own (Smythies, cited by Schaller, op. cit.) and male cubs leave the family at one year. Lion cubs on the other hand, show no signs of independence before they are one year old and male cubs do not usually leave the pride until they are two years old (Guggisberg, 1961).

Whilst cubs are often left by the adults by night (and occasionally by day) the cubs themselves normally stay together. The two females of the thirteen month old cubs were seen a little over 1 km. away from the rest of the pride; this was the youngest any cubs were seen at such a distance from the other cubs of their pride.

Illustrations

PLATE I

Sample of file card showing method of recognition

1a, b, No: C12, Name: Calef, Sex: Male, Age: Born: March 1968, Pride: Romola

PLATE 2

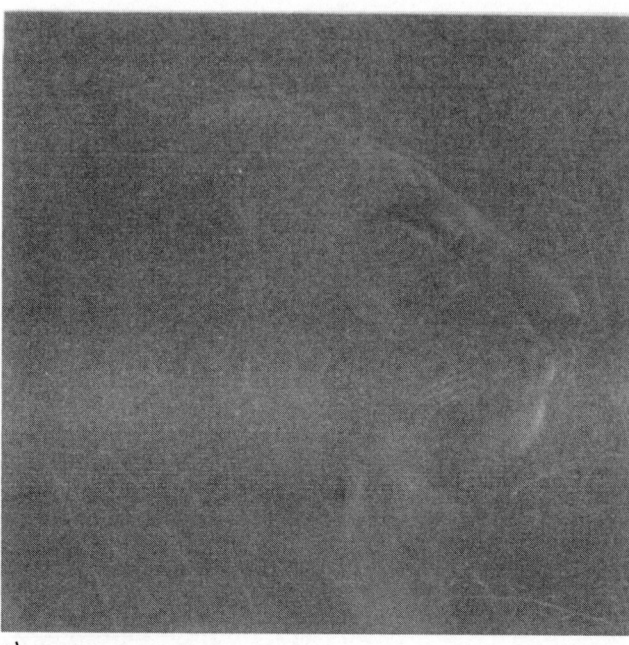

2a

2b

Two photographs of the same animal as a young male (2a) and again two years later (2b). Note that the pattern of spots on the muzzle remains the same demonstrating the validity of this method of identification.

2c Irregular spot between rows 'b' and 'c'

2d Irregular line 'b'

PLATE 3

3a Solid spots on cubs' forehead

3b Rosettes on cubs' body

3c Young male and sister (about 18 months old). Note juvenile spots on female only

3d Mother of 18 month old cubs still showing pronounced spotting

PLATE 4

4a Forepaw with claws sheathed

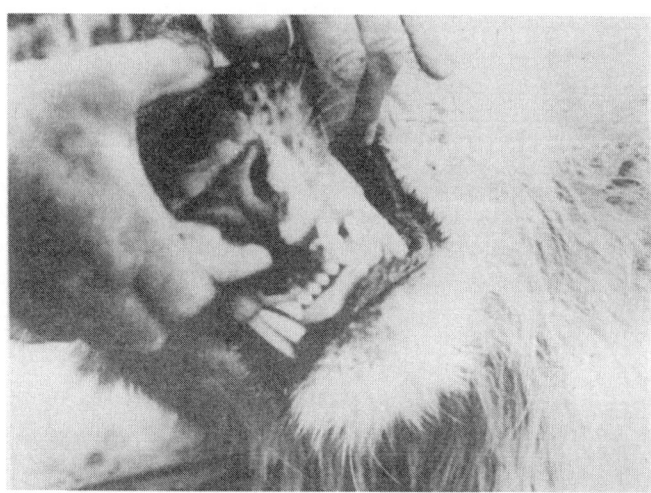

4c Canines of young adult male while under anaesthesia. (Note longer upper canines).

4b Claws extended gripping trunk while cub is climbing tree

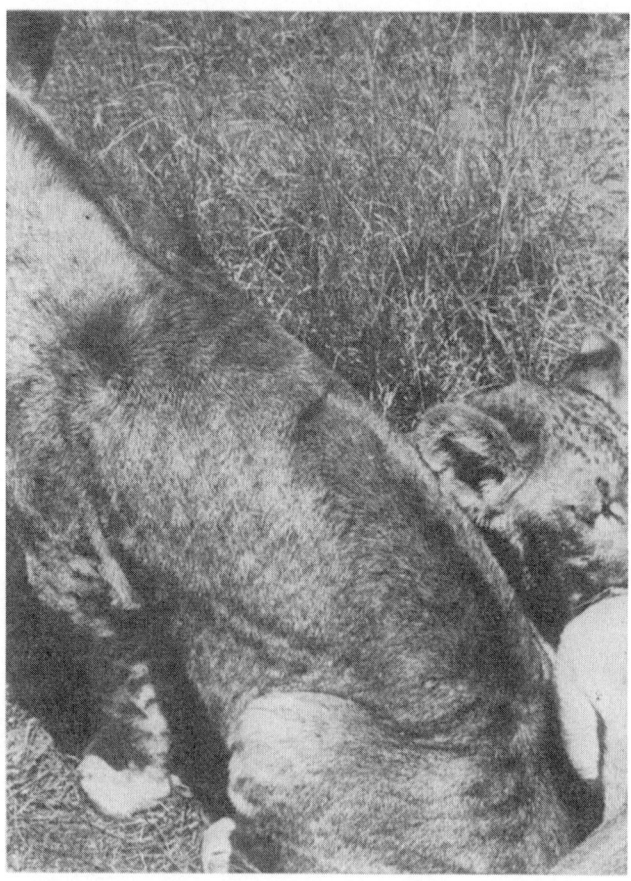

4d Reverse patch and whorl on left shoulder

PLATE 5

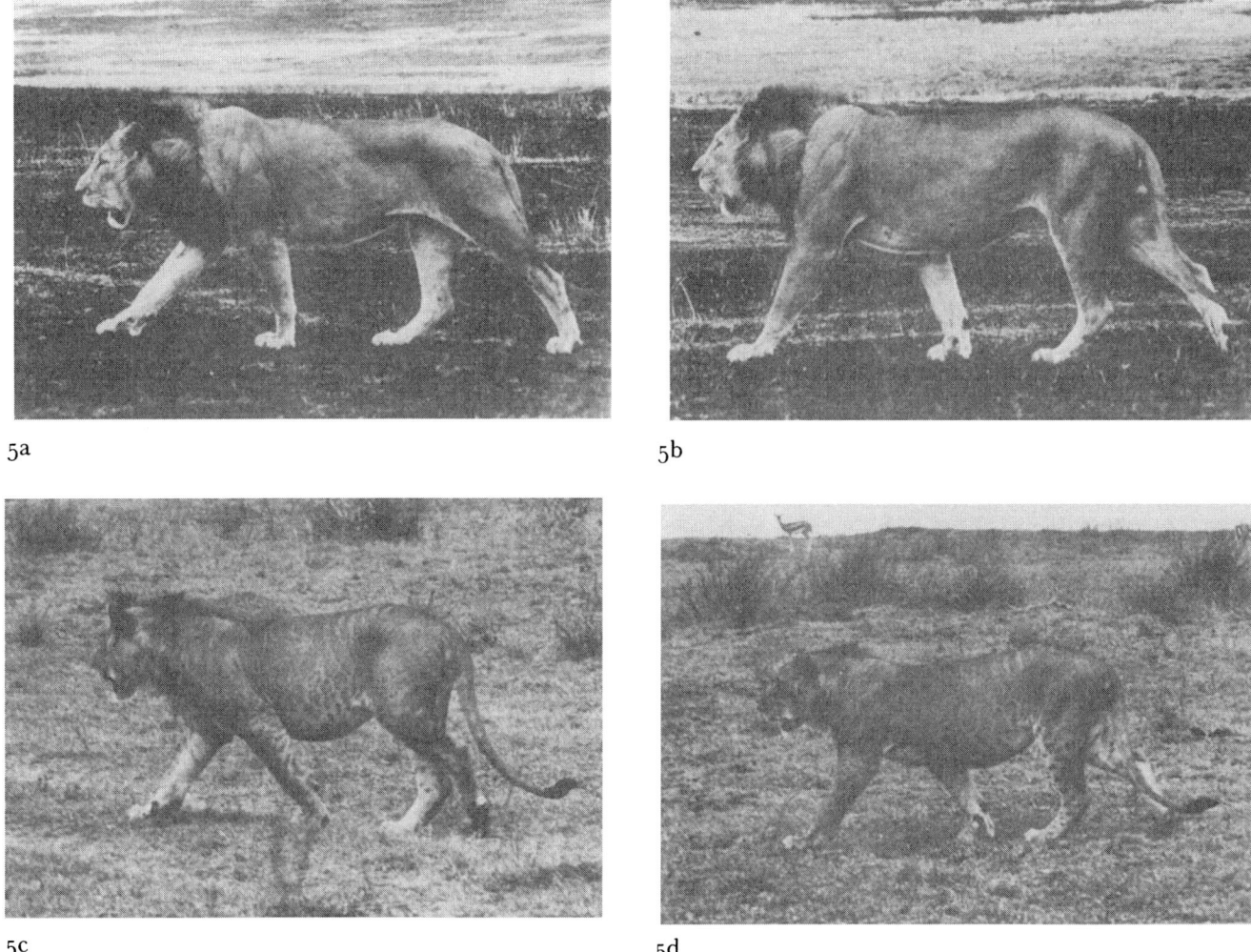

5a

5b

5c

5d

5a, 5b, Pacing walk of adult male
5c & 5d Pacing walk of male cub

PLATE 6

6a   Scar behind ear of lioness

6b   Heavy infestation with ticks

6c   Urinating stance of lioness

6d   Male urinating; downward directed even flow

PLATE 7

7a Male grooming front of mane *Photograph by Mr L. J. Parker*

7b Lioness grooming face

7c Cub grooming front paw

PLATE 8

8a Lioness stretching claws on trunk of Acacia tree   8b Tail flicking

8c Stretching (type 2)

PLATE 9

9a

9b

9a & 9b Young male and sister grooming each other

PLATE 9

9c Cub grooming another

9d Three lionesses grooming one another

PLATE 10

10a Two lionesses greeting

10b Two lionesses greeting

10c Two lionesses greeting

10d Lioness greeted after rejoining pride (Note uplifted tails)

PLATE 11

11a Threat display; snarling

11b Male threatening cub; who ducks with head pressed to ground

11c Male snarling at cub. (Note retracted lips exposing canines)

PLATE 12

12a

12b

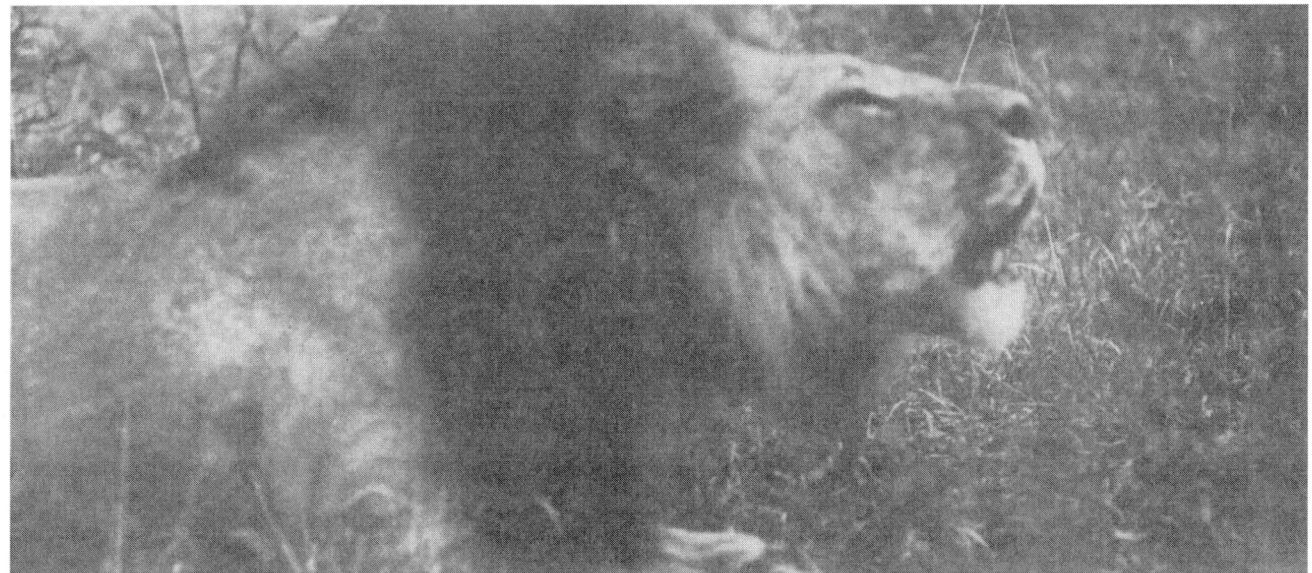

12c

PLATE 13

13a Male marking tree along road where he was travelling. Rubbing head against branches, and 13b subsequently turning around and spraying same area.

13b

13c Female spraying small bush with uplifted tail

13d Male spraying Acacia dreapanolobium tree during mating

PLATE 14

14a Mating pair resting close together

14b "Mating snarl" directed towards lioness

14c "Mating snarl"

PLATE 15

15a Mating pair. Male licking female after she failed to react to mating snarl

15b Male pushing lioness down into crouch

15c Female walking away closely followed by her mate

PLATE 16

16a Copulation with male 'biting' female's head

16b 'Biting'. Note teeth resting on head lightly and lack of reaction to 'bite' from lioness

16c 'Neckbite'

PLATE 17

17a

17b

17c

17d

17a, 17b, 17c, 17d Mating pairs; lioness circling male

PLATE 18

18a Post-copulatory roll by lioness, male standing

18b Female lying on side, male still standing

18c Mating pair surrounded by lioness and cubs, pride of mating female

PLATE 19

19a Cubs playing; 'boxing'

19b Cubs playing; 'boxing'

19c Solitary play; pawing loose rock

PLATE 20

20a Cubs playing in Acacia tree

20b Cubs sitting close together after mother went off hunting

20c Size of approximately two-months-old cub

PLATE 21

21a Two young lions gazing at prey animals. (Note intent gaze, ears directed forward, body in alert position)

21b Lioness stalking zebra, using embankment along road as cover

21c Stalking lioness. (Note body held low to ground, head stretched forward, ears directed forward but held sideways

PLATE 22

22a Digging in warthog burrow

22b Lioness inspects burrow after digging

22c Kongoni alert, keeping lions in sight (Lions are not hunting)

22d Male alerted by vultures overhead, watching them

PLATE 23

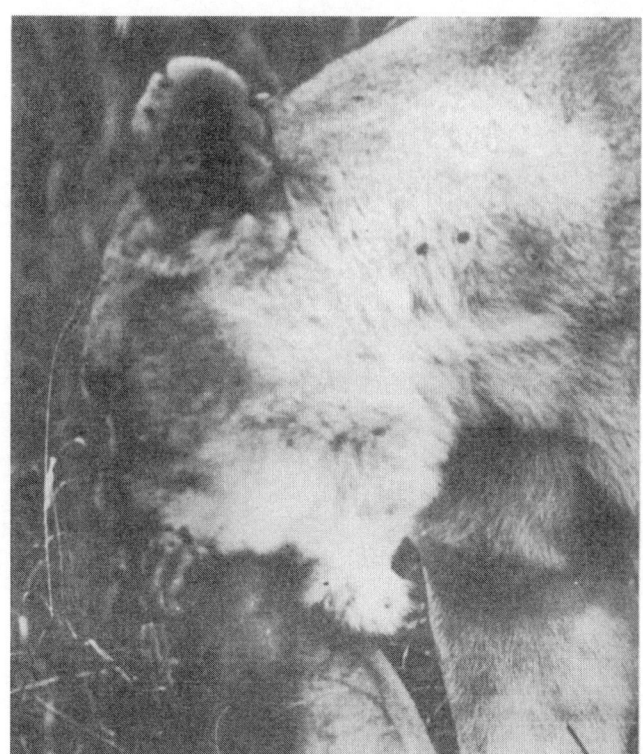

23a & 23b Lioness (Lassie) killing kongoni by 'muzzle-hold' (April 20, 1969)

23c Lassie dragging carcase without releasing hold on muzzle

23d Lioness killing warthog by holding on to throat (strangulation) *Photograph by Mr. L. J. Parker*

PLATE 24

24a Lion dragging carcase

24b Lioness dragging kill

24c Lion chewing zebra skin

PLATE 25

25a Feeding lioness tearing hole in skin with incisors

25b Lioness eating part of small intestine of freshly killed wildebeest

25c Rumen of kongoni left uneaten

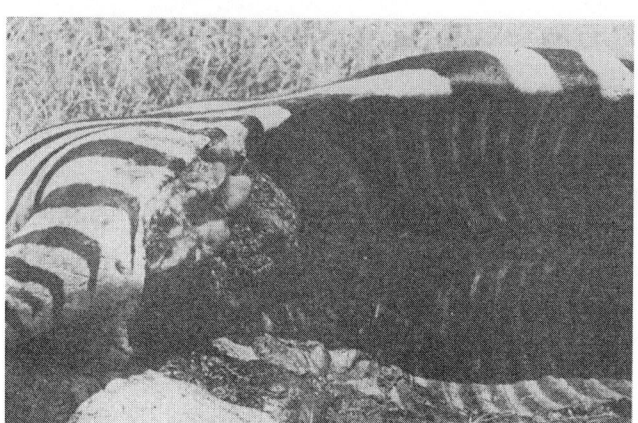

25d Hollowed out carcase of zebra after first stage of feeding

PLATE 26

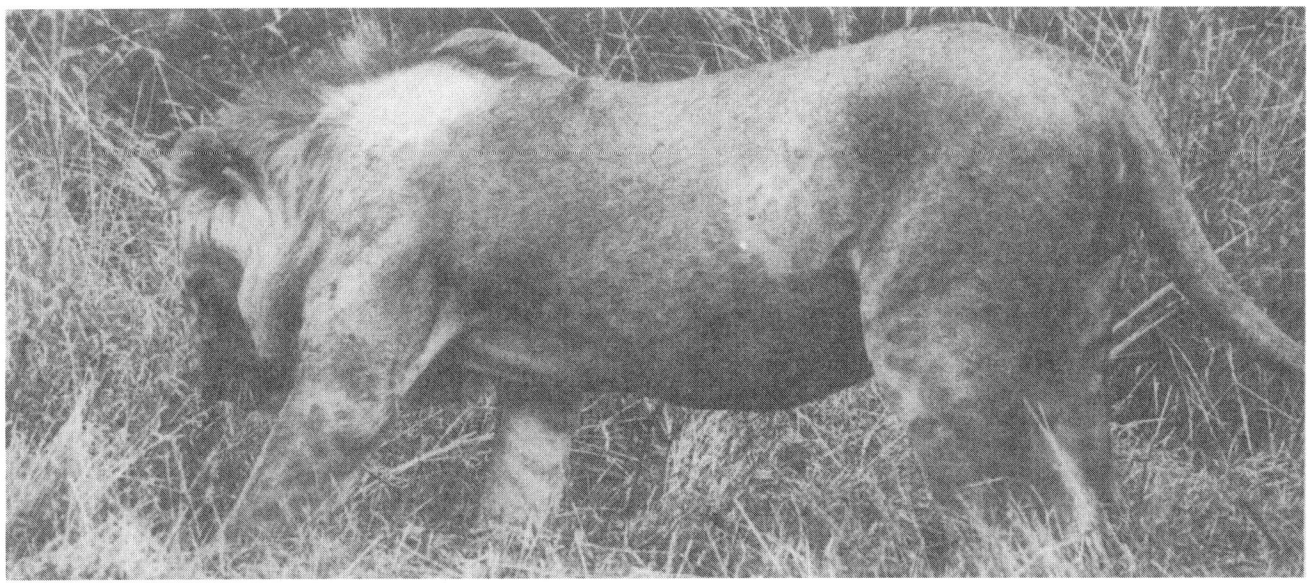

26a

26b

26c
26a, 26b & 26c Scraping near remains of carcase

PLATE 27

27a Drinking position

27b

27c Two female cubs (9 & 11 months old) in very poor condition

27d Male lion after meal (Note extended abdomen)

VII Predation

PREDATION PATTERNS
The prey animals; numbers, density

In any discussion of a predator the prey-predator relationship is of central interest. In this relationship the numbers, kind and habits of the prey species play an important role, as well as the vulnerability of the prey, the latter being influenced to a great extent by the features of the environment (Wright, 1960).

In Nairobi Park it is fortunate that the number of prey animals has been known almost continuously since 1960 through monthly game counts and this, together with the exact knowledge of the lion population for the duration of this study and a fair approximation of their numbers for the previous years, gives more reliable data for a longer period of time than is usually available in this type of study.

Foster and Kearney (1967) compared the 1966 numbers of the most important ungulates in the Park with their numbers since 1961 and came to the conclusion that the population has remained fairly stable since 1962 at about 4000 animals (Foster and Coe, 1968). Table VII–1 lists the present numbers of the most important ungulates and prey species of lion (Column I). The numbers represent an average of twelve game counts during the study period, except in the case of certain animals, indicated in the table, where maximum figures are known to be a more accurate indication of the numbers present. Weights were calculated according to Foster and Coe (1968) and the biomass per km.^{2} entered in Column III. When compared to the figures of Foster and Coe, Column IV, the present data shows a 30% increase in kongoni, continuing an upward trend evident since 1962 (Foster and Coe, op. cit.), a 24% decline in zebra and no appreciable change in wildebeest. The total for the fourteen species listed is very similar to the 1966 figure, thus the biomass seems to have stabilized around the 1962 level.

During the catastrophic drought in 1960–61 the biomass had increased from the previous maximum of 8257 kg/km.^{2} (Petrides, 1956, quoted in Bourliere, 1965) to a new peak of 13215 kg/km.^{2} (Ellis, quoted by Bourliere, 1965), as animals from surrounding areas converged on the sources of permanent water within the Park. Some of the dams were artificially replenished by the Park management and the Embakasi river itself never dried up completely. During the following year (1962) the population was drastically reduced due to lack of nourishment caused by severe overgrazing.

If we take the two figures for the pre-drought period, quoted by Petrides, (minimum: 2180, maximum 8257 kg/km.^{2}) as signifying that a great seasonal fluctuation occurred, then the average of the two would give about 5000 which is only slightly higher than the present average of 4400. This may mean that the Park is now at its pre-drought level, although the composition of the ungulate population has undergone at least two major changes since then: wildebeest have greatly decreased and kongoni have become much more numerous.

The exclusion of domestic livestock in 1967 does not seem to have affected the wild ungulates, as their numbers did not show any change in the rate of increase, as might have been expected after removal of competition. Even the kongoni, expected by Foster and Coe to be the first to take advantage of the new situation, only continued the previous upward trend without acceleration (Fig. VII–1).

Prey–predator ratios

Bourliere (1965) found that the ratio of lions to available prey numbers was fairly constant in Tarangire, Ngorongoro, Rwanda and Eastern Congo at about 1:300. A much higher figure is given by Dasmann and Mossman (1962, quoted by Hirst, 1965) for Wankie National Park in Rhodesia 1:100. The Nairobi figure is closer to this higher ratio at 1:126. However, the number of ungulates per km.^{2} or the stocking density of the prey is comparable in Rwanda and Nairobi at thirty and thirty-two respectively, so that the same prey population supports a higher number of lions in Nairobi.

As noted above, Petrides' figures for 1953–54 suggest about the same biomass as at present, so we may assume that the numbers of prey are also roughly

Table VII–1
Number and
biomass of the most
important prey
species in
Nairobi Park

Species	I Average number	II Percentage of sp. in pop.	III kg/km.^{2} This study	IV kg/km.^{2} Foster & Coe. (1968)
Kongoni	1425	38·5	1,690·0	1,303·8
Wildebeest	259	7·0	373·8	378·0
Zebra	369	10·0	766·4	1,020·3
Warthog	155	4·2	79·6	77·9
Grant's g	419	11·3	184·0	217·4
Thomson's g	232	6·2	39·4	59·0
Waterbuck	64	1·7	83·4	117·8
Impala	449	12·1	177·5	251·7
Bushbuck	32*	0·9	15·0	11·9
Giraffe	98	2·6	658·0	511·0
Eland	47	1·3	149·5	185·9
Reedbuck	10*	0·3	3·6	4·4
Buffalo	20*	0·6	87·0	74·4
Ostrich	124	3·3	122·6	92·8
Total	3,703	100·0	4,429·8	4,306·3

I Average number based on 12 game counts June 1968 to July 1969.
II Percentage of species in population of prey spp. listed on this table.
III Biomass per km.^{2} during this study.
IV Biomass per km.^{2} as quoted by Foster & Coe. (1968) for 1966.

* Maximum figures instead of average figures were used in cases where this is known to represent a more close approximation to the true numbers. Bushbuck and Reedbuck are both non-migratory but elusive animals. Buffalo had been introduced into the Park and the herd is known to consist of at least 20 individuals.

Fig. VII–1

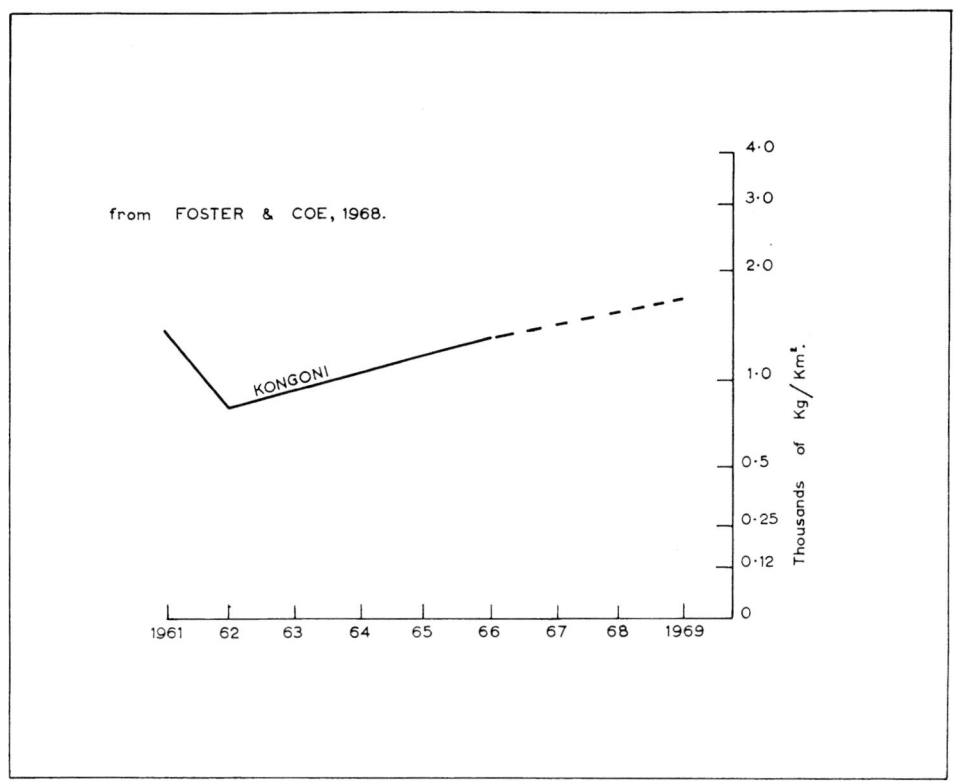

from FOSTER & COE, 1968.

KONGONI

comparable. Although there are no exact records of the number of lions resident in the Park before 1967 we have a figure of twenty-five for 1967 (Foster and Kearney, 1967) which is said to be average for the years 1961–1967. Stephen Ellis, Park Warden from 1952 to 1964, stated (pers. com.) that the maximum number he knew to be resident was thirty-five. The present population of twenty-seven is within the range of these two figures. Thus the lion population since 1952 has also remained substantially stable between twenty-five and thirty-five. As this seemingly high lion to ungulate ratio, as compared with the Tarangire and Congo figures, seems nevertheless to support a stable lion population, it must be assumed that the low lion population in the other areas is not limited by the number of prey available but by other factors, possibly the vulnerability of prey.

Table VII–2 Frequency of various prey species in lion kills in Nairobi and other areas

Species	This study No.	This study per cent	Wright (1960) No.	Wright (1960) per cent	Kruuk & Turner (1967) No.	Kruuk & Turner (1967) per cent	Pienaar (1969) No.	Pienaar (1969) per cent	Mitchell (1965) No.	Mitchell (1965) per cent	Foster & Kearney (1967) No.	Foster & Kearney (1967) per cent
Zebra	17	14·6	—	15	—	26		15·9	30	7·3	—	26·5
Wildebeest	29	25·0	—	49	—	49	—	23·7	25	6·1	—	26·5
Eland	2	1·7	—	2	—	—		0·5	12	2·9	—	1·5
Giraffe	2	1·7	—	4	—	—		4·0	—	—	—	4·4
Kongoni	40	34·5	—	2	—	3	—	—	67	16·4	—	19·2
Warthog	8	6·9	—	—	—	—		1·9	39	9·5	—	8·8
Ostrich	3	2·6	—	trace	—	5		—	—	—	—	5·9
Grant's g.	6	5·2	—	trace	—	5		—	—	—	—	—
Thomson's g.	1	0·9	—	10	—	5		—	—	—	—	—
Impala	1	0·9	—	3	—	—		19·8	—	—	—	5·9
Buffalo	—	—	—	5	—	8		9·3	125	30·5	—	—
Other	7	6·0	—	10	—	—		24·9	112	27·3	—	1·5
Total	116	100·0	150	100	39	101·0	12,223	100·0	410	100·0	68	100·2

Prey selection (based on carcase analysis)

A record of 116 kills is available for the study period. Whenever lions were seen feeding, the skull of the prey was collected for sexing and ageing. In a few instances the skull could not be found, thus some age and sex records are incomplete. Unless it was a known case of scavenging from other predators or on a dead animal, the prey animals were assumed to have been killed by the lions.

Lions utilize a wide spectrum of prey species, there is hardly any edible source of protein that has not been listed at some time as food for lion. Pienaar (1969) lists thirty-eight species as recorded prey, Mitchell et al. (1965) mention nineteen. Although lions as well as tigers (Schaller, 1966; Krumbiegel, 1952) will feed on any vertebrate or invertebrate on occasion, the bulk of their sustenance is provided by mammals, mostly ungulates. The Nairobi Park lions were recorded as preying on fifteen species of mammals and two species of birds, and were observed feeding on ostrich eggs as well.

Although the lions were seen on several occasions to take grass and leaves, a habit shared by other predators such as puma, wolf and tiger (Schaller, 1966; Guggisberg, 1961; Young and Goldman, 1946; Robinette, 1961; Mech, 1966), these are not considered as food. Vegetable matter may possibly act as a scour (Mech, 1966) or as an aid to digestion (Guggisberg, 1961). The grasses eaten by lions were identified as *Panicum coloratum* and *Setaria plicatilis*.

Table VII–2 shows the analysis of kills by species with relative percentages as compared with figures from other sources. Besides the animals listed, waterbuck, reedbuck, hare, steinbuck, marabou stork, monkey (or baboon) and cattle (from outside the Park) were fed upon. The most important species preyed on were kongoni, wildebeest, zebra and warthog in that order. This is in accordance with Kruuk and Turner's contention (1967) that the medium-sized animals play the most important role in the lions' diet although large animals are also taken. Of the 'large' animals listed by Kruuk and Turner only eland and giraffe were killed in Nairobi Park during this study. Although there is a sizeable herd of buffalo, they were not preyed upon. This herd is not truly wild as all individuals were either raised in the Animal Orphanage and then released, or were born to animals thus released. It seems from instances of personal observation, as well as from reliable accounts, that these animals have no fear of lions and instead of avoiding the predators or showing defensive behaviour, they show aggressive behaviour towards them.

Table VII–3
Lion kills according to size (adult prey only)

Large	Kruuk & Turner (1967) No.	Per cent	This study No.	Per cent
Eland	—	—	1	1·4
Giraffe	—	—	—	—
Buffalo	3	8	—	—
Medium				
Wildebeest	19	49	22	30·5
Zebra	10	26	7	9·7
Kongoni	1	3	31	43·0
Ostrich	2	5	3	4·2
Small				
Grant's g	2	5	4	5·6
Impala	—	—	1	1·4
Warthog	—	—	2	2·8
Thomson's g	2	5	1	1·4
Total	39	101	72	100·0

Table VII-4
Preference ratio
based on frequency
of species in kills
and in whole prey
population

	I Per cent in kill	II Per cent in pop.	I/II This study	I/II Foster & Kearney (1967)	I/II Pienaar (1969)
Wildebeest	25·0	7·0	3·6	4·2	3·06
Kongoni	34·5	38·5	0·9	0·7	—
Zebra	14·6	10·0	1·5	2·2	1·98
Warthog	6·9	4·2	1·6	2·2	—
Waterbuck	trace	1·7	—	—	6·05

In contrast to this state of affairs, when a wild buffalo entered the Park in 1962, when there were no resident buffalo, it was killed within 24 hours.

Table VII-3 groups the kills according to size, comparing Nairobi and Serengeti (Kruuk and Turner, 1967) figures. (Only adult prey were taken into consideration in this table). There is a preponderance of medium prey in both. Most observers agree (Pienaar, 1969; Foster and Kearney, 1967) that the number of small prey is under-estimated in these studies, although the degree of the bias depends on the individual circumstances. It is possible that the higher number of small prey in this study is due to the fact that Nairobi Park is a small area and fewer kills were thus missed than in Serengeti.

The most conspicuous fact emerging from Table VII-2 is the high predation on the small wildebeest population. This has been noted by Foster and Kearney (op. cit.) who showed that wildebeest have been captured since 1954 at a greater rate than expected on the basis of their numbers in the population, even while this species was abundant. In 1961 this rate was more than twice the expected, while in 1966 it was four times the expected. Table VII-4 gives percentage in kills and percentage in population for wildebeest, kongoni, zebra and warthog and provides figures comparable to Foster and Kearney's and Pienaar's calculations. (The figures for percentage in population were based on yearly averages from game counts; see Table VII-1).

The figure in the third column in Table VII-4 represents the ratio of the two percentages, i.e. percentage in kill over percentage in population, called the 'preference ratio' by Pienaar (1969). The preference ratio for wildebeest at 3·6 shows a slight decline from the figure calculated on the basis of data given by Foster and Kearney, while the total number of wildebeest remained substantially stable (mean for 1966: 252, for this study: 259). Pienaar (1969) also reports a preferential selection of wildebeest in Kruger Park of comparable degree. There it is exceeded only by waterbuck, which is almost twice as high, and by kudu. Hirst (1969) in his study in a Transvaal area also noted a preference for wildebeest and suggested similar habitat as a possible cause. Although no figures on preference ratio were given, he concluded that a preferential selection did exist.

While wildebeest has slightly declined in preference since 1966, kongoni underwent a slight increase in this respect, keeping pace with the 30% increase in their relative abundance. The ratio of preference for 1966 was 0·6, but at the end of this study it was 0·89.

Of the three most important prey species zebra alone shows a definite decrease in numbers. There has been a 24% decline in the population since 1966. During the same period its preference ratio declined from 2·5 to 1·45.

Preferential selection of warthog revealed a slight decrease while the population remained substantially stable.

These figures suggest that prey selection may very gradually become adjusted to the distribution of the species in the population as in all four cases preference ratios are closer to '1' for 1968-69 than for 1966. However, the basic preference for wildebeest still exists. They were killed at twice the expected rate when abundant in 1961; the discrepancy increased after the sharp decline of the population and seems now to be slowly swinging back towards the lower figure.

Such adjustments to changing conditions do not always occur: in Kruger Park when buffalo increased, so did predation on the species, but the increase in zebra numbers was not followed by a similar increase in mortality due to predation (Pienaar, 1969).

It is interesting to compare the food habits of the Nairobi lions with those in other areas. Preferential food habits have been noted by many authors and the seeming neglect of some abundant species is a frequent occurrence, while the

Table VII–5
Sex distribution in lion kills

Species	Male		Female		Unknown		Total	
	No.	Per cent	No.	Per cent	No.	Per cent	No.	Per cent
Wildebeest	9	17·3	16	50·0	4	12·5	29	25·0
Kongoni	27	52·0	6	18·8	7	21·9	40	34·5
Zebra	8	15·4	4	12·5	5	15·6	17	14·6
Warthog	—	—	—	—	8	25·0	8	6·9
Ostrich	3	5·8	—	—	—	—	3	2·6
Eland	2	3·8	—	—	—	—	2	1·7
Giraffe	—	—	1	3·1	1	3·1	2	1·7
Grant's g	1	1·9	4	12·5	1	3·1	6	5·2
Thomson's g	1	1·9	—	—	—	—	1	0·9
Impala	—	—	1	3·1	—	—	1	0·9
Other	1	1·9	—	—	6	18·8	7	6·0
Total	52	100·0	32	100·0	32	100·0	116	100·0

very same prey animal may form the preferred food of the lions' diet in another area. Thus in Kruger Park Pienaar (1969) noted a marked preference (pref. ratio over six) for waterbuck even in areas where wildebeest were abundant, although the latter still had a fairly high preference rating. This is difficult to explain on the basis of the waterbuck inhabiting favourable habitat from the viewpoint of the lions' hunting success (riverine bush), since at the same time impala in Kruger Park, inhabiting similar country, are low on the preference list even though abundant. Similarly puku are very little preyed upon in Kafue, though present in great numbers (Mitchell et al., 1965). In contrast to Kruger Park, waterbuck have not been preyed upon to any appreciable extent in Nairobi (less than 1% of prey) although they form 2% of the population.

The very low figure for Thomson's gazelle in the kills in Nairobi contrast with Wright's findings (1960), where 10% of kills was made up by this small gazelle. The fact that he combined figures from Nairobi with those obtained in other areas, including the Serengeti Plains, may account for this. During certain seasons this gazelle makes up a large percentage of the lion kills in Serengeti due to scarcity of other available prey (Kruuk and Turner, 1967).

Table VII–5 shows all kills tabulated according to sex. Sexing and ageing was on the basis of horn shape in wildebeest and kongoni (Watson and Gosling,

Table VII–6
Sex ratio in lion kills and population for wildebeest and kongoni

Species	M. to F. ratio in pop.*	M. to F. ratio in kills	x^2	Probability
Wildebeest	1 : 1·60	1 : 1·78	0·18	Over 0·5
Kongoni	1 : 1·59	1 : 0·22	23·23	Less than 0·001

* From game count figures.

pers. com.) and on the basis of teeth in zebra (Klingel, 1966). Warthog age was estimated on the basis of size and tusk size according to Roth (1965). No attempt was made to sex warthog.

An overall selection for males is evident from these figures in the proportion of 62% m. to 38% f. This is almost identical with the 60:40 ratio found by Wright for all predators and 59:41 for lions only. Very similar figures are given by Mitchell (1965); 56:44 and Hirst (1969); 57:43.

If these figures in the kill are compared with the sex ratio in the various species, we find that selection for sex is not uniform. In wildebeest, for instance, there does not seem to be any selection as the sex ratio for the population is almost identical with that in the kill (Table VII–6). While in Nairobi more female than male wildebeest were killed. Hirst (1969) and Pienaar (1969) mention preferential selection for males. Without knowing the sex ratio of the population in their areas, their data cannot be compared with the conclusions drawn here.

In kongoni kills we find a strong selection for males. The difference in sex ratio in the population and in kills is statistically highly significant (Table VII–6).

Walther (1969) states that during his study of the Thomson's gazelle more males than females were killed, as adult males have a shorter flight distance than females. In contrast to Estes & Goddard (1967), who linked a similar higher mortality in males with territoriality, Walther found that very few territorial males were killed. He maintains that it is the bachelor males who take the brunt of predation, partly due to their being forced into the less favourable habitats as far as safety from predation is concerned. Gosling (pers. com.) tends to agree with this view, maintaining that in his opinion it is those males that are non-territorial, or have lost their territories and are too old to regain them, that make up the bulk of the prey in kongoni. This view finds additional support in the fact that more than 20% of the kongoni kills were classified as 'old'.

The same lowered flight distance of males as found in Thomson's gazelle may underline the general preponderance of males in most of the kill data analysed (Mitchell et al., 1965; Wright, 1960; Hirst, 1969; Makacha & Schaller, 1969). Mitchell agrees with this explanation; however he then goes on to say that excess mortality of males will be reflected in the sex ratio of the population. But it is doubtful whether predation on any one species is high enough to influence the sex ratio in the population.

While it used to be believed that predators were effective culling agents by removing mostly aged and diseased individuals, some of the recent predation studies cast doubt upon these suppositions (Wright, 1960; Stenlund, 1955).

Table VII–7
Age distribution in
lion kills

Species	Young No.	Per cent	Subad. No.	Per cent	Adult No.	Per cent	Old No.	Per cent	Unknown No.	Per cent	Total No.	Per cent
Wildebeest	4	26·7	3	17·6	21	33·9	1	10·0	—	—	29	25·0
Kongoni	5	33·3	2	11·8	24	38·8	7	70·0	2	16·7	40	34·5
Zebra	4	26·7	5	29·4	6	9·7	1	10·0	1	8·3	17	14·6
Warthog	1	6·7	4	23·5	2	3·2	—	—	1	8·3	8	6·9
Ostrich	—	—	—	—	3	4·8	—	—	—	—	3	2·6
Eland	—	—	1	5·9	1	1·6	—	—	—	—	2	1·7
Giraffe	—	—	2	11·8	—	—	—	—	—	—	2	1·7
Grant's g.	1	6·7	—	—	3	4·8	1	10·0	1	8·3	6	5·2
Thomson's g.	—	—	—	—	1	1·6	—	—	—	—	1	0·9
Impala	—	—	—	—	1	1·6	—	—	—	—	1	0·9
Other	—	—	—	—	—	—	—	—	7	58·4	7	6·0
Total	15	100·1	17	100·0	62	100·0	10	100·0	12	100·0	116	100·0

They show that in many cases adults, seemingly in their prime both in age and in physical condition, are sustaining the heaviest losses.

Kühme (1964) and Wright (1960) state that most of the lion kills were of prime age and in Wright's survey 93% were in good health. However, it seems that a considerable proportion of wolf kills is made up of sick, old or crippled individuals (Murie, 1944; Cowan, 1947; Crisler, 1956); similarly most puma kills showed some handicapping condition (Hibley in Errington, 1946).

In this study 62 out of 104 kills aged, that is 60%, were classified as adults (Table VII-7). But again there are considerable differences among the various prey species. Although for both kongoni and wildebeest adult kills outnumbered all other classes combined, in zebra the situation was reversed; adults constituted only 37·5% of kills. This accords with Wright's findings for this species, indicating that adult zebra are less vulnerable than adults of wildebeest or kongoni.

No assessment was made during this study of the physical condition of animals killed. As very little of a lion kill is left besides bones, it was felt that no valid assessment could be made. It is quite possible that, as Cowan (quoted in Mech, 1966) proposed, a high proportion of prey classified as in prime condition may nevertheless harbour heavy parasitic infestations, such as the hydatid cysts (Echinococcus) he found in elk kills. As Sachs and Debbie (1969) state, it is not known what degree of infestation by parasites will cause symptoms, in other words, what degree would make an animal more vulnerable to predation. Thus this whole question of selection according to physical condition is difficult to assess at present.

Mitchell et al. (op. cit.) found a strong seasonal trend in the lion kills in the Kafue Park, with more kills made during the dry than during the wet season, while they detected no such difference in the leopard kills analysed. The monthly figures in this study are too low for statistical analysis. To divide the study period into dry and wet seasons also proved impracticable, as the rainy season was erratic and precipitation too widespread, thus no sharp division could be made between rainy and dry seasons.

Habitat as a factor in prey selection

Table VII-8 shows all kills according to the type of habitat where they were made. While it is true that lions may drag a kill away before starting to feed on it, due to the size of the three main species examined they would not be dragged very far in most cases. In many instances the spot where the actual kill was made could be verified and thus the error in this classification reduced to, it is hoped, a negligible figure.

The most interesting finding in this table is the very high percentage of wildebeest kills made in *Acacia drepanolobium* areas as opposed to open grassland, the wildebeest's normal habitat. It might have been expected that most wildebeest kills would be made in the open plains and in fact that is what

Table VII-8
Location of kills according to habitat types

Habitat type	Wildebeest		Kongoni		Zebra		Other		Total	
	No.	Per cent	No.	Per cent	No.	Per cent	No.	Per cent	No.	Per cent
Grassland (2)	6	20·7	9	22·5	4	23·5	7	23·3	26	22·4
Scattered bush (6)	3	10·3	11	27·5	7	41·2	9	30·0	30	25·9
Acacia scrub (1)	13	44·8	15	37·5	6	35·3	7	23·3	41	35·3
Riverine forest (5)	7	24·2	3	7·5	—	—	1	3·3	11	9·5
Other or unknown	—	—	2	5·0	—	—	6	20·0	8	6·9
Total	29	100·0	40	100·0	17	100·0	30	99·9	116	100·0

Figures in parenthesis refer to Verdcourt's classification as quoted in Appendix I.

Kruuk and Turner's (1967) figures show; 74% in open plain. However, in the present study only 20% of wildebeest kills were made in the open grassland and almost half in *Acacia* scrub. Another 24% were killed in the riverine forest. The majority of tiger kills found by Schaller were made in riverbeds (Schaller, 1967).

The location of most of the *Acacia* area kills makes it a possibility that the animals were on their way to water, as they were located within a few hundred metres of riverbeds. The kills in riverine thicket were almost certainly made when the wildebeest were going to drink. A very low proportion of kills was made in scattered bush country, a type of habitat not often frequented by this animal.

These figures may give one possible explanation for the high frequency with which wildebeest kills are made out of a small population. It may be that the extensive *Acacia* scrub area is a very favourable one for the lions' hunting technique; while it provides cover for the stalk it does not impede the last rush towards the prey. While scattered bush area would give similar advantages it is usually avoided by wildebeest. Although by day wildebeest are rarely found deep in the *Acacia* areas, they are often on the edges and seem to enter them more often at night, or during the very early morning hours, presumably in many instances traversing these stretches of *Acacia* on their way to water.

Thus it seems that the presence of the *Acacia* scrub in Nairobi Park is a contributing factor to the high predation on the small population of this ungulate. A basic preference for wildebeest evidenced by various findings in different areas (Hirst, 1969; Pienaar, 1969), combined with the large areas of habitat more favourable to the predator than the customary open plains where wildebeest are usually found, keep the preference ratio on this animal at a high level.

Another noteworthy fact is the absence of zebra kills in riverine bush. It seems that as they have to stand in water in order to drink, they seldom use the steepsided rivercourses, but prefer to drink from the dams which are located in open plains, have shallow edges and little, if any, fringing vegetation.

Weir et al. (1965), during a study of game animals visiting a waterhole, found that the peak drinking hours for wildebeest were between 0500 and 0900 hours while for zebra the peak hour was at 1800. In fact, judging from the condition of carcases when found, most wildebeest kills were made during these early hours, and on the basis of circumstances described above, most seem to be made when the animals went for water. Zebra on the other hand seemed to be rarely killed when going to water; this may be due to a combination of their selection of open drinking places and the time of their drinking in the early evening just before the start of the lions' hunting activities.

HUNTING METHODS AND FEEDING BEHAVIOUR

Searching

It has often been debated whether cats hunt by sight or by smell. As Grzimek's experiments in the Serengeti with lions and zebra dummies have demonstrated, optical orientation is the primary response to the presence of potential prey and is then verified by the sense of smell (Grzimek, 1960; Schaller, 1967). Leyhausen, on the other hand, attributes no role to the sense of smell in locating prey by domestic cats (1956 a).

Hunting can be divided into the following behaviour sequences; searching, approaching, catching, killing and feeding (Leyhausen, op. cit.).

Searching for prey appears to be done mostly by night by the Nairobi lion population, as most of the daylight hours are spent resting and apparently not in active search of prey. Yet, whenever a potential prey animal appeared within the visual field of the lions, even though they were resting or even feeding, they showed some interest by looking in the direction of the animal, sitting or standing up, taking a few steps or making a definite attempt at approaching the prey (Kühme, 1965) (Plate 21 a). This approach may sometimes be just an unconcealed trotting towards the animal; it is mostly cubs that behave in this manner, but occasionally adults, especially young

Fig. VII–2a

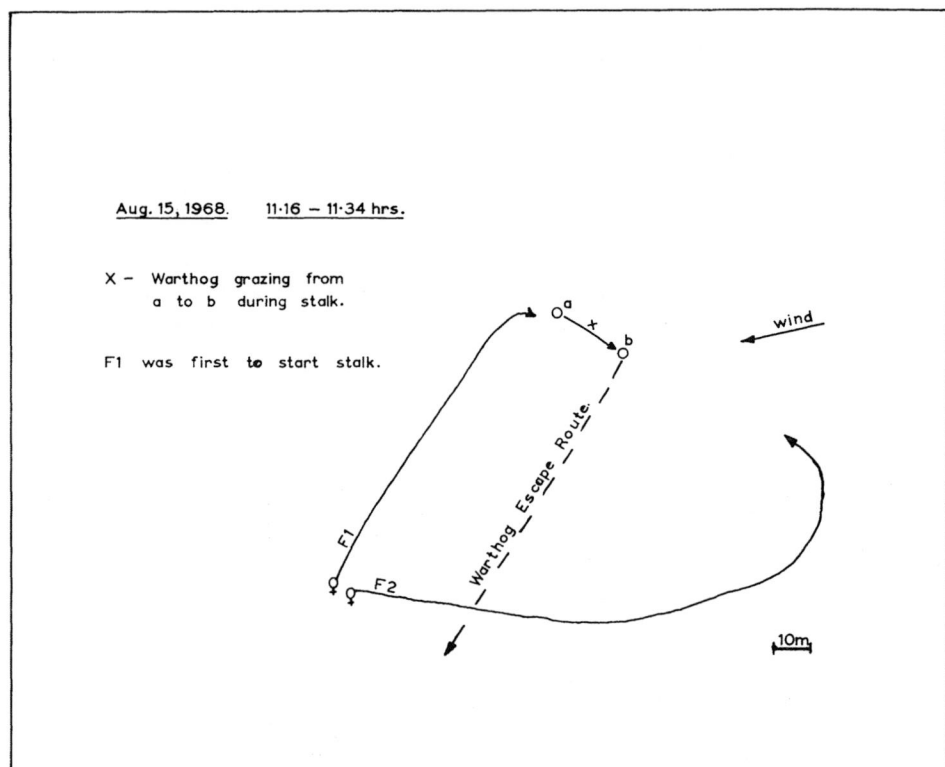

Fig. VII–2b

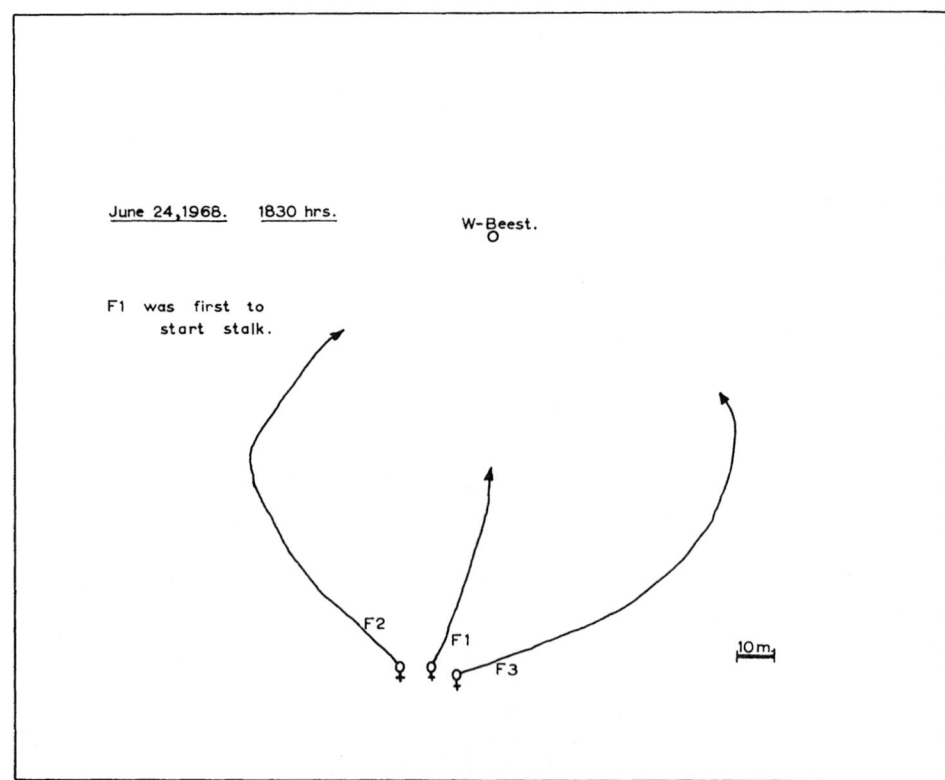

lionesses, and on one occasion even the adult male was seen to approach a herd of grazing animals in this way. Such obviously ineffective chases are usually given up after less than fifty metres and the animal then returns to his previous position.

As the intensity of response to potential prey varies with the degree of hunger of the predator (Precht, 1952), a more efficient pursuit will follow the initial reaction when an animal has not fed recently. The usual approach used by lions is similar to the stalking method used by other felids (Leyhausen, op. cit.) and even some canids like the wild dog, although the latter, unlike the lion, then follows up with a long chase (Estes and Goddard, 1967). Stalking is characterised by walking with body held low, head stretched forward, ears directed forward; the whole body appears tense (Plate 21 c). Steps may be taken quickly or very slowly, with the gaze fixed onto the prey. A typical feature of this stalking locomotion is the frequent 'freezing' of the predator, sometimes with a paw in mid-air, whenever the prey seems to direct its attention towards the hunting animal. Between stalking sequences the cats often adopt a crouching position, taking advantage of any cover available, such as termite mounds, clumps of vegetation, banks along the road, or slopes in the terrain (Plate 21 b). A sloping terrain may afford good cover and may compensate for lack of any other cover in an open short-grass and otherwise bare area. One lioness was observed to crouch and stalk from behind a standing bus; the lions in Kruger Park often use cars in a similar manner (Pienaar, 1968).

It had been thought that lions take the direction of the wind into account when hunting, but Schaller (1967, 1969) maintains that neither lions nor tigers pay any attention to it whatsoever.

In all the stalks observed during this study, when direction of wind was noted, it was either blowing towards the lions or across the line between prey and predator. In no case was a stalk started when the lions were upwind from their intended prey, as in that case the ungulates would probably be alerted even before the lions would have a chance to circle around to the downwind side.

The first lion to start a stalk usually sets off in a straight line towards the prey. The only time this was not the case, a sub-adult lioness started a stalk by circling to the far side of a warthog family, even though there was a crosswind. When a second animal joins in the stalk it will take a course at an angle to the first and if there are three, they will fan out and thus attempt to encircle the prey (Kühme, 1966) (Fig. VII-2). Occasionally an animal was seen to approach moving prey by anticipating its movements.

In contrast to cheetah which can outrun many of the ungulates within a short distance (Schaller, 1968; Kruuk and Turner, 1967), a lion, like a tiger, has to get very close to its intended victim before a successful final rush can be made. Schaller gives this distance as 10–25 m. for tigers and it seems to be similar for lions. In one case two lionesses had approached a warthog to within 25 and 50 m. respectively (Fig. VII-2 a) but the warthog was still able to break away. However, in this case, as well as in several other instances, one gained the impression that the predator could have been successful if it had put more effort into the pursuit. But it often seemed as in this case, that a lion was unwilling to expend maximum energy on these attempts. It is possible that on these occasions a sufficiently strong motivation was lacking, even though in the case just cited eighteen minutes were spent on the stalk.

Preference ratio based on stalking

Earlier in this chapter I have shown that according to a preference ratio based on actual kills wildebeest was the most preferred species. I have

Table VII–9
Preference ratio
based on frequency
of stalks

Species	I Stalks		II Per cent in pop.	Per cent I/II Pref. ratio
	No.	Per cent		
Wildebeest	6	9·8	7·0	1·4
Kongoni	19	31·2	38·5	0·8
Zebra	4	6·6	10·0	0·7
Warthog	12	19·7	4·2	4·7
Ostrich	3	4·9	3·3	1·5
Eland	1	1·6	1·3	1·3
Other or undetermined	16	26·2		
Total	61	100·0		

attempted a similar analysis for preference on the basis of stalks observed. Table VII–9 gives a 'preference ratio' on the basis of 61 observed stalks, calculated in the same manner as the one in Table VII–4. In arriving at these results open chases, as described on Page 86 were not included but only stalks with the characteristics described in paragraph 2 of Page 87. Forty-one stalks were timed, they ranged from one to fifty-three minutes, with an average of 12·7 minutes.

It might have been assumed that the large percentage of wildebeest in the actual kills resulted from a larger success in hunting it. However, the figures on stalks confirm that there is already an initial preferential selection for this animal as against kongoni or zebra.

The highest preference rating on this basis, however, is for warthog. In fact, it was often observed that when there were several species of grazing animals within sight, the lions concentrated their attention on warthog to the exclusion of others. This, of course, holds true only by day, as warthogs spend the night in their burrows.

For kongoni this latter preference ratio is the same as the ratio based on kills; for zebra it is less than the ratio based on kills.

It was formerly believed that predators and scavengers were two easily distinguishable types of carnivores. However, many recent studies (Kruuk, 1965; Estes, 1967; Walther, 1969) prove that this is not so. In fact, the roles traditionally assigned to some carnivores can, under certain circumstances, be reversed as with hyaenas and lions in Ngorongoro (Kruuk, 1966; Schaller, 1969; Kühme, 1966). During this study three definite instances of scavenging by lion were recorded. In some other cases there were indications that an animal which was fed upon had not been killed by the lions but this was not conclusively proved. In one of the three known cases a Grant's gazelle was scavenged from a female cheetah (with four cubs) which had made a kill a few hundred metres from the resting lions. The cheetah abandoned the carcase as soon as the pride approached (McLaughlin, pers. com.). In the second a dead Grant's gazelle was found and eaten by the male lion. The cause of death was not known but the carcase was intact when found. In the third case Scarface was resting when he suddenly stood up, looked into the distance, then started to trot, occasionally slowing down to a walk, then trot again and finally gallop and charge into a group of vultures assembled on a freshly stripped kongoni carcase. He had covered about 1 km. in a straight line in about 20 minutes. After picking up the carcase and carrying it 10 m. away under a small *Acacia drepanolobium* tree, he proceeded to feed on the remains. The most probable explanation for this behaviour is that the lion had noticed vultures descending over the spot, although I did not detect them. However, on several other occasions lions were seen watching attentively as vultures circled overhead (Plate 22 d) and Kruuk and Turner (1967) made similar observations in the Serengeti Park where lions were scavenging from vultures after seeing them alight near a carcase. Kühme (1966) and Schaller (1969) also reported cases of lions scavenging from other predators.

Attacking and killing

Much has been said about the methods of attacking and killing prey by lions. The various methods of killing described by earlier authors have given rise to controversies about which is 'the' method by which lions dispose of their prey. But as more systematic observations were made it became evident that the method varies with the prey, the environment, and other circumstances, taking into account the number of lions attacking. E. Young, (1966). Guggisberg (1961) and Eloff (1964) claim that lions attack their victim from the front, while Grzimek (1960) and Kruuk and Turner (1967) contend that they attack from behind. Tigers may attack from either direction (Schaller, 1966) while Felinae approach at an angle from behind (Leyhausen, 1965 b).

While in primitive carnivores indiscriminate biting functions for both catching and killing, in the higher viverrids and felids the forelimbs have become more and more specialized for seizing and the jaws and teeth for killing (Leyhausen, op. cit.). In some Felinae a specialized killing bite has evolved and become oriented towards the neck of the victim. This 'neck-shape taxis' is innate in some felids and viverrids. In this respect one can speak of 'the' killing bite for a cat (Leyhausen, op. cit.) while in lions the method of killing is more flexible.

While the bite in the victim's nape has variously been reported for Pantherinae, both lions and tigers (Guggisberg, 1961; Selous in Guggisberg, 1961; Schaller, 1969), a more frequently used method seems to be killing by strangulation or suffocation (Guggisberg, pers. com.; Schaller, pers. com.; Kruuk and Turner, 1967). Strangulation is also used by cheetah (Schaller, 1968) and tiger (Schaller, 1966). Strangulation is induced by holding on to the throat and suffocation by clamping down on the muzzle, covering and closing the nostrils and the mouth.

In the two instances when the attacking and killing of a large animal was witnessed during this study, the identical method of suffocation by holding the muzzle was used by two different lionesses, Romola and Lassie. A single lioness was in pursuit of a wildebeest in one case and a kongoni in the other. In each case the lioness approached from behind at an acute angle, jumped onto the animal's back and grabbed the muzzle, presumably while the victim was turning its head in an attempt to dislodge the attacker. Both lion and prey then sank to the ground together, the lioness holding on to the animal's body with her claws and keeping her mouth-hold over the muzzle. There was very little struggling in either case. In the case of the wildebeest the lioness held on for about one-and-a-half minutes before she let go and stood up. When the antelope then made some slight movement with its legs, she took hold of the muzzle again and held it for another minute and forty seconds and then released her hold. There was no further movement.

In the case of the kongoni the sequence was similar, except that the kongoni fell to the ground together with the lioness more quickly than the wildebeest. The lioness jumped onto the lumbar region of the hartebeest, which caused the hindquarters to collapse immediately with the lioness hanging on to the muzzle with her mouth (Plate 23 a, b). The kongoni kicked only a few times and was apparently dead in less than sixty seconds. The lioness held on for two to three minutes after it stopped moving and then started dragging the carcase towards a thicket without changing her hold on the muzzle (Plate 23 c). In neither case did the lioness change her hold after the first grab, as cats sometimes do (Leyhausen, 1965 b), nor was any other part of the animal bitten.

In the case of a reedbuck and a Grant's gazelle kill when several lionesses were involved, the lionesses were all holding on to the animal's neck region, pressing it down to the ground, but none was holding the muzzle. In both cases only the last stage of seizing and killing the prey was witnessed, so it is not known whether all attacked the animal together, or in succession, or in what way.

When an adult warthog was caught by three females, the exact moment of grabbing was not seen, but a few seconds after the actual seizing one female was holding the warthog by the throat while it was lying on its back kicking and another lioness held its muzzle. It seems that suffocation by the muzzle-hold does not work well in the case of warthog as the tusks prevent the lion from closing its mouth over the muzzle effectively to shut out all air. The lioness holding on to the muzzle in this case was a young female and she let go shortly, while the older lioness was holding on to the throat for over five minutes before releasing her hold. The warthog struggled for a longer time than either the wildebeest or the kongoni, as possibly the stranglehold on the thick neck of a pig is not as efficient as suffocation by the muzzle-hold (Plate 23 d).

From the examples witnessed, the muzzle-hold seems an extremely effective method, involving very little struggling by the victim and thus little danger to the attacker, provided the first grip is well placed. It has the advantages of the hold onto the throat cited by Schaller (1967) such as safety from horns and hooves, with the added advantage of being quicker.

Prof. R. Hoffman, late Head of the Anatomy Department of the Veterinary School, University College, Nairobi, confirmed my impression that the hold over the nose would cause suffocation more quickly than the stranglehold on the throat. The soft parts of the nose and the slit nostrils of the ungulate make it possible for the predator to compress the vestibulum completely. With the muzzle-hold, air is thus effectively prevented from entering either through the nostrils or through the mouth. The trachea, on the other hand, is fairly stiff and more difficult to press together completely so as to shut out all air.

This killing technique is a very specialized method as it is indirect: not killing by the pressure of teeth, i.e. actual biting, but by a combined holding and pressure action of the jaws and possibly the tongue.

It is interesting that the hold over the nose has not so far been described for any other felid, but Mech (1966) repeatedly saw wolves attach themselves to the nose of moose, while other members of the pack attacked various parts of the body. Mech states that the hold on the nose is not mortal but only distracting to the victim thus facilitating the overpowering of this large herbivore by the much smaller predators. However, it is not impossible that in some instances, when the wolf held on long enough—in one case this hold was maintained for 10 min.—suffocation was the cause of death. The killing suffocating hold of the lion may have developed from a similar beginning, when among several lions attacking, one held on to the nose to distract the victim, or to prevent its using horns for defence. Like the strangulating hold on the throat, the muzzle-hold is an adaptation for dealing with large prey (Schaller, 1967).

Smaller prey may be grabbed anywhere (Guggisberg, 1961); often it may be killed by the mere grabbing (Schaller, pers. com.) but this is not always the case. With small prey the lions do not make any effort to kill before starting to feed. The function of killing is purely to render the victim defenceless, or prevent it from escaping. Cats are well known for 'playing' with their prey (Leyhausen 1965 b) before killing it, although they would be perfectly able to kill it outright. In one instance a baby kongoni was caught by three lionesses, then appropriated by the male who started to feed on it. While he was alternately feeding and leaving the kongoni in order to charge other members of the pride in defence of his prey, the calf repeatedly ran away and was caught again and again. The lion held it down with his paw, then grabbed it by the neck or head and carried it back to his original position to continue feeding on it. No attempt seemed to be made to kill it outright, although this could easily have been done. One hour after it was first caught, the calf was still heard to bleat. However, this repeated catching of the prey without killing did not show any of the elements described by Leyhausen (op. cit.) as 'playing with prey'.

In several cases during this study more than one animal was killed within a short time. In the case of the wildebeest kill described above, another wildebeest, apparently killed a short time earlier, was discovered lying about 400 m. from where the second kill was made. The second carcase was subsequently dragged near to the first. As the first carcase was inaccessible, it could not be seen whether it was still intact and thus could have been killed immediately before the second, or whether some of it had already been consumed. In several other cases more than one animal was killed; two female wildebeest and a new-born; a zebra with her young and on two separate occasions a female wildebeest and her young. In all these instances the pride stayed on the kill until all animals were consumed, although there are cases mentioned where lions killed several animals but fed only on some of them. Pienaar (1969) mentions a case where fifteen buffalo were killed by a pride, but only a few were fed upon.

Leyhausen's experiments with cats (1965 b) showed that the prey-catching and killing co-ordinations are separate and specifically exhaustible. Thus, it is possible that, under certain circumstances, catching or killing may become appetitive rather than consummatory behaviour. Such a situation may explain instances of multiple killings referred to above.

Another method of catching prey is the digging out of warthogs from their burrow (Plate 22 a, b). On one occasion a female was seen digging unsuccessfully at a warthog burrow and another time two young males dug out a warthog from its burrow and killed it. Similar behaviour was reported by Guggisberg (1961), Pienaar (1969) and Mitchell et al. (1965).

Feeding

In most instances, after the prey has been killed, it is carried away some distance before feeding is started. Felids eat their prey on the spot only when the kill is a very small animal (Leyhausen, 1956 a), otherwise they carry it about for a while and then take it to a place sheltered from view and light. In lions both these behaviour elements have been observed. In instances where considerations of light or shelter did not apply, the kill was nevertheless moved from its original position before feeding was started. Scarface was observed to behave in this way on two occasions, when he moved kills made at night for a few metres before settling down to feed. By day, the kill was usually dragged or carried a distance to a more sheltered, or at least a shady, position. On one of the occasions mentioned above the male was feeding all night on a carcase in the open (after having moved it, but not into a sheltered place) and then at 0600 hours he carried it into dense bush. Thus carcases were often moved into dense cover in the morning and brought out again into the open after sunset. Similar behaviour was described in tigers (Schaller, 1967).

Thus we are dealing here with two different behaviour elements: one, the carrying about before feeding, regardless of a sheltered or open position; and the other, the hiding of the carcase from view and/or light.

Displacement of the carcase can be achieved by carrying, dragging or pulling. Small prey is usually carried by holding it by the neck or throat. When the animal is too big to be carried, it is dragged between the legs. This seems a laborious process and requires considerable effort. It is usually done

Total time dragged	Total time rested	Total time	Lion	Prey
Seconds	Seconds	Seconds	Sex	Species
30	30	60	Female	Kongoni
251 (13 stages)	529	780	Female	Kongoni
110 (4 stages)	70	180	Male	Wildebeest

in small stages with frequent rest periods. Carcases were dragged for distances ranging up to 400 m., with an average of about 50 m. Small animals, which can be carried without dragging, can of course be taken much further. The 400 m. distance was observed when a lioness dragged a half-eaten carcase of a kongoni during a period of 50 min. An adult zebra was dragged by a lioness for about 50 m. and an adult wildebeest was dragged and carried by Scarface for approximately 50 m. Periods of dragging between rests range from 2 sec. to 36 sec. for lionesses (thirty-eight observations) with an average of 15 sec. and from 10 sec. to 80 sec. for the male with an average of thirty-five (five observations). The greater strength and endurance of the male is also seen in the table above showing the relative length of dragging and rest stages.

Times were adjusted to the nearest full second and distances were estimated, except in the case of the 400 m. distance where a map reading was taken.

When an animal was too large or heavy to be dragged between the legs, the lion turned around and pulled it while walking backwards.

There is an interval between the death of the prey and the time the predator that made the kill starts feeding. As in cats (Leyhausen, 1956 a) which start feeding immediately only on small prey, in all observed cases the lioness which killed the animal started feeding only after an interval of variable length, although other members of the pride often started feeding immediately. When Lassie killed a kongoni, she first lay panting next to it and only started feeding 26 min. later, after having dragged it 30 m. into the thicket. When Romola killed a wildebeest, she did not seem exhausted, yet started feeding only 20 min. later. Again when a warthog was killed, all cubs and Romola started feeding immediately, but not Patricia, who was the one holding on to the throat. She walked about, moaned softly, seemed agitated, disappeared and returned, played and only started feeding 60 min. after killing the pig. This walking about is similar to the 'inspection' of the surroundings mentioned by Leyhausen (1956), even when the cat knows the area well. In contrast to this behaviour, tigers start feeding immediately, if the carcase is in a suitable location (Schaller, 1967). Kühme (1965) gives an account of a lion which scavenged a carcase from wild dogs and did not start feeding on it until 15 min. later.

As Leyhausen has pointed out (1956 b), the lying position for feeding is a distinguishing characteristic of the Pantherinae among the felids and so is their use of their forepaws to hold the prey while feeding on it. Felinae eat mostly in a crouched position, only very rarely lying on their haunches. The tiger lies, crouches, stands or sits for feeding and uses its forepaws but little (Schaller, 1967). Lions may occasionally stand or sit for short periods, but were never observed to feed while crouching.

The carcase is most often opened on the underbelly between the hindlegs. Less frequently feeding starts on the rump. It is interesting that the male was always seen starting on the rump, which is also where tigers start feeding (Schaller, op. cit.) and the females, with few exceptions, by opening the belly. In two cases a female did start on the rump, but almost immediately went over to open the belly and then proceeded as usual. This may serve to expose the soft inner organs, rich in nutrients, to the cubs, who cannot break through the skin of the larger herbivores unaided.

No animal was eaten head first as is the habit of Felinae (Leyhausen, 1956 a), not even small prey where such a procedure would be feasible.

To open a carcase the skin may be pulled up by the incisors until a small hole is made, then the carnassials are used to cut it open (Plate 25 a). When feeding starts on the underside the intestines are first pulled out, a fact also noted by Guggisberg (1961) and some usually eaten (Plate 25 b). A lioness was observed eating the small intestines of a wildebeest, pulling the guts along between her teeth first as if to squeeze out the contents. However, the soft or liquid contents of the intestines may occasionally be consumed while the pellets of formed faeces are rejected. The stomach and its contents and the contents of the large intestine and sometimes part of the guts themselves are left uneaten (Plate 25 c). Tigers often leave the intestines and never consume the contents of the digestive tract (Schaller, op. cit.). An attempt is usually made to cover the stomach, the guts or their contents. This is done by repeated and alternate scraping motions with the forepaws, one to twelve sweeps with each paw (Plate 26 a, b, c). In this manner loose soil, leaves or grass may be swept over the remains and sometimes the pawing is so vigorous that it up-roots surrounding grass. Often, however, this scraping is performed in vacuo; it may be carried out either on the spot where the carcase lay before it was dragged away, or a lioness may return repeatedly to the carcase to perform these scraping sweeps of her paw near it, with nothing visible to cover. Similar 'symbolic' covering motions were seen by Guggisberg (op.cit.) near the stomach of a prey but he states that he rarely saw a lion actually cover up remains.

In most of the cases observed all the inner organs were consumed first. Both Cooper (1942) and Krumbiegel (1952) noted a preference for internal organs in captive lions. In one exceptional case the intestines of an adult eland were left in the body cavity untouched for unknown reasons; in two other cases part of the lungs and the liver were left. The usual method however was to hollow out the carcase first (Plate 25 d).

After the inner organs, the hindquarters were usually eaten starting on the rump, then the forequarters; the head being left until the last.

Fig. VII–3

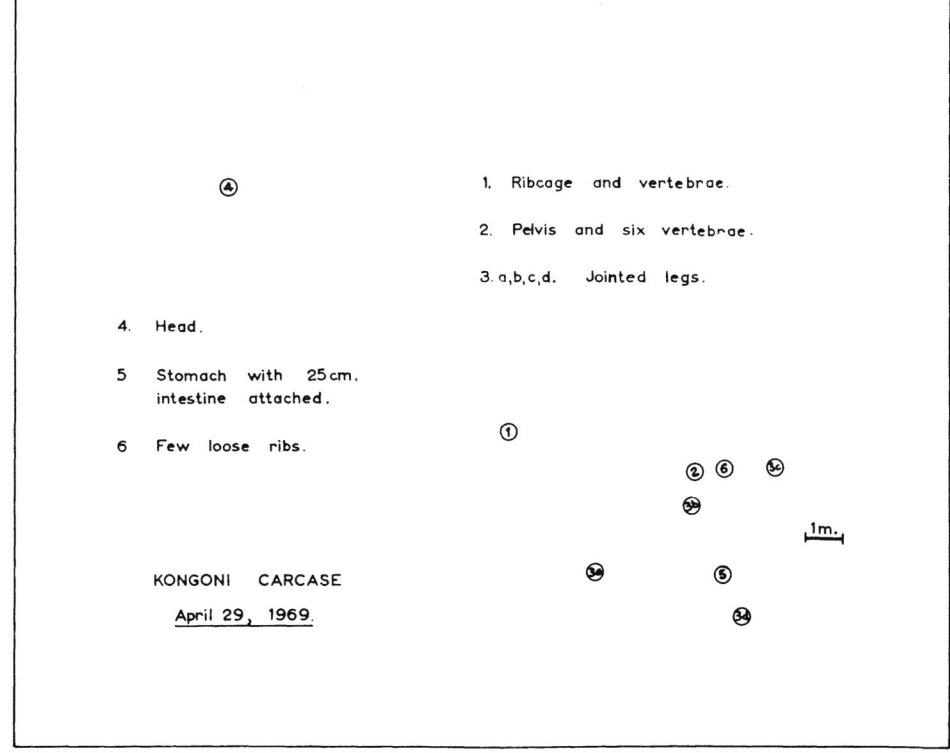

1. Ribcage and vertebrae.

2. Pelvis and six vertebrae.

3. a,b,c,d. Jointed legs.

4. Head.

5 Stomach with 25 cm. intestine attached.

6 Few loose ribs.

KONGONI CARCASE
April 29, 1969.

1m.

All blood running out when the prey was opened was carefully licked up, hardly any was ever wasted. This licking up of the blood may have given rise to the habit of extensive and frequent licking of the carcase, even the unbroken skin, where no blood is evident.

This behaviour was also noted by Cooper (1942) in captive lions, but there the flesh itself was licked, as only butchered meat was given. The skin is usually completely consumed and, as a rule, only the stripped bones are left. The rough tongue of the lion, with its backward-directed papillae can very efficiently scrape the meat from the bones. The skeleton is usually left disjointed in the following manner: the skull with a few cervical vertebrae attached; the rest of the vertebral column with ribs attached; pelvis; legs disjointed with lower two joints intact with skin and hooves. While the tiger may devour the horny part of the hooves, the lion only does this with very young or very small animals, when the whole carcase is consumed with only a few bone splinters, if any, left.

While in very young animals all bones are eaten including teeth and hooves, in adults only the edges of ribs and scapula are gnawed and the nasal bone is occasionally chewed off; the other bones are left intact.

The remains of a carcase are usually found within a small area, but as the head is the last to be stripped, sometimes the pride may move off with one animal carrying the head along after all the rest has been consumed. Thus a cub was seen carrying the head of a sub-adult kongoni for over 1 km. before disappearing in a thicket.

While lions were never seen to shake their prey after seizing it, not even small animals, they were often observed to shake the carcase, or parts of it, while feeding. Shaking of prey is a killing technique in some canids, while in some felids its function seems to be to stun the reflexes and reduce struggling of the victim while it is carried by the predator (Leyhausen, 1965).

Both Guggisberg (1961) and Kruuk and Turner (1967), state that lions often return to a kill after they have left the carcase. On only two occasions was such behaviour observed, when a pride returned in the afternoon to a carcase they left that morning. In one case they did not return to feed, only rested about 50 m. from the kill; in another they were seen feeding again on the remains. They never attempted to cover a carcase when they left it, as tigers and puma are known to do (Schaller, 1967; Young and Goldman, 1946). In this study lions were seen to cover up only those parts which are never consumed. This behaviour, however, must have developed for the purpose of covering up the uneaten carcase to protect it during the predator's temporary absence. In the case of the lion this function is no longer important as it is very rare for lions to leave a partly eaten carcase unattended only to return to feed on it later (only once seen during this study). With lions hunting and feeding communally, the original function of this behaviour is now almost meaningless and is evident only in the covering up of unwanted remains as described above, or is performed in vacuo. It may be a behavioural relic from ancestors, that had to leave half-eaten carcases temporarily, or it may be part of the behavioural repertoire of the genus which in this species has no function but which may be important in a related species. This may be a similar case to territorial marking, which is non-functional in the non-territorial zebra, but meaningful in related territorial species (Klingel, 1967).

Consumption; kill frequency

An attempt was made to establish how much one animal eats from any given prey, but the calculations must be regarded as very tentative, since in the case of large prey it was never possible to observe a pride uninterruptedly from the time the kill was made until it was left and so one cannot know whether other lions than those seen, had shared in the kill. When small prey was caught and consumed, the figures obtained do not indicate the full capacity of a lion's consumption.

A hare was eaten by a lioness in 8 min. and a few weeks old kongoni was consumed by a male in 80 min. leaving only bone splinters. A lioness finished a steinbuck weighing approximately 13 kg. (Maberley, 1965) in 33 min., leaving only one hoof. It took 4 hours for two lionesses and five 10 and 12 month-old cubs to finish feeding on an adult warthog. In this case the remains were weighed and the amount eaten calculated as about 46 kg. (total weight according to Lamprey in Foster and Coe, 1968). This represents a 78% utilization, as against 64% reported for a waterbuffalo killed by a tiger (Schaller, 1966). The pride thus consumed about 6·5 kg. per animal, considerably less than a lion is capable of eating in one session (Pienaar, 1969; Schaller, 1969).

An adult kongoni's remains were also weighed: the meat consumed was about 125 kg. representing approximately 75% utilization, very similar to the warthog carcase. (Total weight according to Stewart in Foster and Coe, 1968) (see Table VII–11 for detailed calculations.) This amount was presumably eaten by a mother with three 5-month-old cubs in 4 days during which they stayed on the kill. Assuming no other lion shared the meal, this would mean a daily consumption of about 31 kg. for an adult and three small cubs.

Table VII–11
Amounts consumed: utilization of carcase

Warthog consumed by 2 lionesses and 5 cubs (10 and 12 months old) Remains in kg.		
Head	6·8	Total average weight according to Foster & Coe
Stomach and contents	1·1	(1968): 59·0 kg.
4 feet	2·8	Approximate amount consumed: 46 kg.
Skeleton	2·7	Utilization: 78%
Total	13·4	

Kongoni consumed by lioness and 3 cubs (5 months old) Remains in kg.		
Head	4·9	Total average weight according to Foster & Coe
Skeleton and skin	19·4	(1968): 165·7 kg.
Rumen and contents	16·9*	Approximate amount consumed: 124·5 kg.
		Utilization 75%
Total	41·2	

* Mean of 18 samples collected by M. S. Price (pers. com.).

A pride of three females, one male and five 4- and 6-month-old cubs stayed on an adult eland kill for 4 days. Assuming similar 80% utilization based on 363 kg. weight (Foster and Coe, op. cit.) about 72 kg. would be consumed per day by the 4 adults and 5 cubs.

In one instance the pride of three lionesses and five 8- and 10-month-old cubs were on an adult zebra for 24 hours at the most. If no other lion had fed on the kill, they consumed about 190 kg. (80% of 238 kg. quoted by Foster and Coe, op. cit.) in less than 24 hours, or some 24 kg. per lion. Schaller (1969) states that five lions can easily devour a 270 kg. zebra in one day,

Table VII–12
Estimates of kills
per lion per year

Source	I Kill lion/yr.	II Applied to Nrbi.	III Per cent of total Nrbi. prey pop. (3,703)
Stevenson-Hamilton in Pienaar, 1969	10–12	324	8·7
Wells in Guggisberg, 1961	15	405	10·8
Guggisberg, 1961	20	540	14·5
Stevenson-Hamilton in Guggisberg, 1961	26	702	18·8
Talbot and Talbot in Pienaar, 1969	35	945	25
Wright, 1960	36·5	985·5	26·4
Pienaar, 1969	15	405	10·8

I Gives estimate of various authors of lion kills per year per lion.

II Applies these estimates to the present Nairobi lion population of 10 adults and 17 juveniles to give total kills per year for the Park.

III Represents the percentage of prey population (as given in Table VII–1) that the lions would eliminate per year.

which amounts to 43 kg. per lion (80% of 270 kg.) or over 20% of the lion's own weight (see Plate 27 d).

The amount of prey killed by lions during a year is a question that has given rise to much speculation and many estimates have been put forward ranging from ten prey animals per lion per year to over thirty-six. These data are summarized in Table VII–12 together with the number of animals that would be killed in Nairobi Park on the basis of each of these estimates, assuming that the figures apply to the total lion population, not only to adult animals. It is also assumed that they apply to medium sized ungulates, the lion's most frequent prey, but only E. F. V. Wells specifies prey weighing about 260 kg. (117 lb.), the approximate size of a kongoni.

So many variables enter into these estimates, however, such as young to adult ratio in the lion population, which in Nairobi seems exceptionally high, and the relative frequency of small, medium and large prey in any given area, that the value of these figures is very problematical.

During this study the longest period for which a pride was kept under observation and all kills made during that time recorded, was an 8-day period during which the Romola pride (three lionesses and five 10-and 12-month-old cubs) made seven kills including warthog, ostrich, zebra, kongoni and wildebeest. That such successful periods of hunting are by no means the rule, can be seen from the photographs on Plate 27 b, c. The condition of the two cubs (about 9- and 11-month-old females) from this same pride proves that they must have gone through a period of very limited hunting success. Guggisberg (1961) also notes that the pattern of killing is very irregular and one or two lean weeks may very well follow one of abundance. Thus I feel no meaningful calculations on kill frequency can be based on this 8-day period as there is no reliable yardstick to indicate the correction factor to apply to this period of abundance to give a realistic estimate. Results from several periods of continuous observation, preferably at different seasons of the year, would be necessary to arrive at a reliable figure for the killing frequency of the Nairobi Park lion population.

Drinking

As in the case of the tigers mentioned by Schaller (1967), lions left a kill immediately after they had finished feeding, even though they often did not travel far before settling down again. Usually they went first to the nearest water for a drink. If they had stayed on a kill for half a day or longer they sometimes even went to drink in between feeding periods. Most instances of drinking were observed in the morning, as, when the prey was small or the pride large, prey caught during the night was usually finished during the morning, between 0700 hours and 1200 hours. The lion drinks in a crouching position (Plate 27 a) similar to the feeding position of Felinae, with the hindquarters often elevated. This drinking position was only once seen in another context, when the male was looking for small scraps of cartilage and skin while finishing a zebra carcase.

Drinking is a lapping action, like in domestic cats and dogs. The tongue is extended with the tip curled under backwards and water is thus lifted up to the mouth.

Drinking periods ranged from a few seconds to over 6 minutes with an average of 2 minutes (forty observations). Lions do not seem to be particular about the kind of water they drink: they will drink from rain-puddles as readily as from flowing or standing bodies of water. A lioness was seen drinking from a stagnant pool with a thick green film over it.

Hunting success

Estes & Goddard (1967) calculated the hunting success of a pack of wild dogs as 85%. This represents a higher success rate than was found in any other predator. Rudebeck (quoted in Mech, 1966) studied the hunting habits of various predators in Europe and found an overall success rate of 7·6%, representing percentage of successful kills out of total attempts. A very similar figure, 7·8%, was arrived at for wolves hunting moose (Mech, 1966).

During this study, six out of 61 stalks recorded, that is roughly 10%, were successful. This is of the same order as the 8% success in tiger stalks observed by Schaller (1967); however, he concluded that the true figure was probably closer to 5%.

As two of these six successful stalks were for warthog, which is considered easier prey than most larger animals (Mitchell et al., 1965) and is moreover not available at night, the true figure would be lower, probably between 5 and 10%; which is in keeping with the figures of Rudebeck and Mech.

VIII Reaction to Environment

REACTION TO HUMANS AND VEHICLES

The Nairobi Park lions are famous for their tolerance of vehicles even in fairly close proximity. However, this tolerance is not complete indifference and in many cases definite reactions of varying intensity to both vehicles and humans can be observed.

Reactions of lions to cars ranged from a slight lifting of the head and looking towards the vehicle to actually hitting it with the paw. Cubs showed more flight reaction than adults and the adult male showed more aggressive behaviour than any other individual.

Reactions occurred to auditory as well as to visual stimuli. A sudden revving of engines or loud human voices heard when unseen road workers were talking, or a car's horn that started blowing continuously caused the lions to look in the direction of the sound. The response depended on the suddenness and intensity of the stimulus and ranged from interest to avoidance reaction in cubs.

Intensity of response to cars depends largely on factors listed by Walther (1969) as influencing flight reaction in Thomson's gazelle:
(1) suddenness
(2) direct approach
(3) size
(4) strangeness
(5) experience of animal
(1) is connected in this context with speed of approaching car; a vehicle approaching at greater speed elicited a greater avoidance reaction from cubs than a slowly approaching one.
(2) driving straight towards the animals caused flight more easily than an indirect approach;
(3) buses caused more pronounced avoidance reactions on the part of cubs than small cars;
(4) humans in unusual positions for the lions caused great interest. Whenever a person was seen above the top of a car without a roof to cover the silhouette, the animals, and especially the cubs, looked intently at it until it disappeared. Walther (op. cit.) found that a person standing in front of a Land Rover caused no reaction, whilst the same person in the same spot and at the same distance from the animals, but standing free against a natural background, caused a strong flight reaction. The same

observation was made in the case of lions; the human figure against a natural background caused the greatest reaction.

(5) young cubs or adults, when first seen in the Park, were more shy than those that had been exposed to encounters with vehicles more often.

The most frequent response to vehicles was just looking in the direction of the car: most adult females showed either this mild interest or none at all. Occasionally a female showed slight threat display, as when Romola was suckling her three-month-old cubs and was disturbed by the sudden revving of an engine: she got up, took a few steps towards the car and snarled (see threat display iv). Cubs were occasionally seen to crouch and make silent snarl faces at cars or run to their mother and huddle close to her. At other times they retreated and sometimes hid, emerging a few minutes later and returning to their previous place.

The most intense reaction to any vehicle or its occupants was shown by Scarface. On 2nd August 1968 whilst mating, Scarface growled and swished his tail when a man showed his head above the roof of a car. He growled and snarled at the passengers in a saloon car on 17th September, and 24th October, both times whilst mating. On both of these occasions the male was passing the car at about 1 m. distance or less, whilst following the female. On 25th October, when still mating, he repeated the same performance and, in addition, hit the side window with his paw twice before walking on. On two other occasions, both times when mating, Scarface lifted his tail and sprayed a stationary vehicle when walking past it.

The most violent reaction by a lioness was seen from Anne, a young adult, who, when first seen in the Park on 15th July 1968, crouched and backed away in a crouching position, when approached by a Landrover. She showed similar behaviour on 28th October 1968, when she backed away from a bus and then turned and ran to the older lioness, Misty, who lay at about 50 m. distance from the bus.

In contrast to their reaction to vehicles in motion or approaching them, a car that has been stationary for some time may be approached by the lions quite closely, investigated, rubbed against and sometimes used as a source of shade. On 10th February 1969, the Romola pride was resting in the open. Around midday, they gradually settled down in the shade of my Landrover as no other shade was available within 100 m.

Kühme (1965 b) noted that the lions in the Serengeti became much more trusting by night and quotes Eibl-Eibesfeld (1950) and Hediger (1934) as confirming that nocturnal animals become bolder at night. It is difficult to determine whether this, in fact, is the case in Nairobi, as the animals showed occasional lack of shyness both by night and by day. But as they were more likely to be on the roads or out in the open and active by night than by day, it is difficult to evaluate the difference, if any, in their behaviour.

Overflying planes were occasionally noticed by the lions, but it could not be determined whether they reacted first to the sound or the sight of the moving planes. Although they showed no fear or flight reaction, cubs were more apt to react to planes than adults. They lifted their heads and followed the movement of the planes with their eyes. This is in keeping with the finding that cubs are more reactive to novel stimuli than adults (Glickman, 1964).

INTERACTION WITH OTHER SPECIES

The reaction of the ungulates that form the usual diet of lions varies with the behaviour of the predators. When lions walk openly or are at rest, the behaviour of the herbivores is usually one of cautious watchfulness rather than panic. (Plate 22 c) Kihara, an adult male, was seen on 7th June 1969, to walk in an open short grass area straight at a herd of wildebeest who scattered at 50 m. distance, then stopped, turned around and stood at about the same distance in a semi-circle around the lion who lay down to rest facing away from the antelopes.

Zebra similarly surround resting predators in a semi-circle observing them with interest, head erect with forward directed ears (Klingel, 1967). Flight distance in this animal was measured by Klingel and found to be 80–100 m. It was often observed by the writer as well as by Klingel that zebra will approach lions, then stop at their usual flight distance and keep watch over them.

Kongoni, wildebeest, Thomson's and Grant's gazelles take flight at between 60 m. and 100 m. from approaching lions, but invariably stop and look back, then repeat this performance several times before finally departing.

Ungulates often give an alarm call when becoming aware of lions and before taking flight. On one occasion when two lionesses were stalking an adult female Grant's gazelle it took flight, stopped at 90 m. distance, looked back at the lionesses and gave an alarm call. The gazelle continued to give the call for the next eleven minutes while slowly walking away, making many stops to look back towards the lionesses.

The alarm of one herbivore often warns other species; thus an alarm call from an impala seemed to alert zebras when Chryse was stalking a herd on 20th July 1969. On another occasion the bark of a troop of baboons alerted a herd of Grant's gazelle at the approach of a pride.

In one instance a lioness and her three cubs left the remains of a kill when a herd of buffalo approached. It is probable that the approach of the herd caused her to leave as instances of buffalo herds attacking and chasing lions and killing cubs by trampling on them have been reported (Makacha & Schaller, 1969; Uganda Game, 1951, quoted by Cullen, 1969; Guggisberg, 1961).

Baboons sometimes line up on a broad front facing and observing the lions from about 100 m. distance, similar to wildebeest and zebra discussed above, but usually barking continuously. On one of these occasions Scarface was lying flat on his side; at the sound of a sharp bark he lifted his head with a growl then lay back again. Baboons in trees often gave their barking alarm call when lions passed below them, but the latter paid no attention to these sounds.

Of the larger herbivores seldom preyed upon, only rhino were seen close enough to the lions to show any interaction. Rhino and lion usually pay no attention to each other and have in fact occasionally been seen resting in each other's company (Guggisberg, op. cit.).

On 17th November 1968, three lionesses, a lion and five 6- and 8-month-old cubs were on a wildebeest kill, the male feeding whilst the others were resting nearby. At 0710 hours all became attentive and looked fixedly towards the west, where an adult rhino was approaching them at about 100 m. distance. The rhino continued coming until at about 20 m. distance it stopped, looking at the pride. Two of the females stood up, the cubs and a young lioness sat up whilst the male lifted his head and all gazed steadily at the rhino which slowly turned his head from Scarface at one end of the line to the females at the other. When the rhino took another few steps towards them, the cubs retreated, followed by the lionesses. The rhino then turned and walked away several steps, immediately followed by Chryse and the cubs, who retreated again as soon as the rhino stopped and turned back, advancing towards them. This was repeated several times with the rhino alternately approaching and retreating and Chryse and the cubs following its movements, advancing and retreating in turn, until at 0717 the rhino finally walked away. Chryse followed its movements with her eyes long after all the others had relaxed. At 0719 Scarface resumed feeding.

On another occasion a rhino turned away from 150 m. distance after sniffing the air, being downwind from two mating lions who apparently did not notice its presence. Another time a rhino walked past a lioness and her three cubs at less than 100 m. with neither paying any attention to the other.

Several authors have pointed out the intolerance of carnivores towards each other (Schaller and Lowther, 1969, Cullen, 1969, Guggisberg, 1961). Lions may kill and are in turn killed—when old or crippled—by hyaenas or wild dogs. They have been known also to kill, and occasionally devour, cheetah and leopard (Kenya Parks, 1957, Downey's Africa, quoted by Cullen, 1969).

The intolerance of lions towards hyaenas was evident on the few occasions when hyaenas were encountered during this study. On 25th March 1969, two lionesses and three cubs were on a wildebeest kill when a hyaena was seen carrying a piece of the carcase away, followed by one of the lionesses in a stalking gait. When Scarface arrived at the scene a few minutes later, he chased the hyaena for 200 m. When he turned back towards the kill, two hyaenas were seen walking back and forth behind him but did not approach the pride again. At the same time two jackals were circling around Scarface at 10 m. distance but he paid no attention to them. Adult lions seldom reacted to jackals, but cubs or young females occasionally chased or stalked them. Jackals often approached lions to within 10 m. and barked at them for several minutes without any reaction from the cats. This often happened when the lions were walking; the jackals ran up to them, then followed them closely, barking all the while. The significance of this behaviour is not understood.

When a tape of barking baboons and calling hyaenas was played to Scarface and the Romola pride, only one animal, a cub, looked towards the sound of the baboon call. The reaction of the male to the hyaena call, however, was intense and immediate. He suddenly sat up, looked in the direction of the call, then trotted towards it, apparently looking for the source of the sound.

When he started moving, the recording was stopped but he continued in the direction from which he had first heard the call and in so doing passed the vehicle containing the recorder. When the tape was played again, he turned back towards the sound and repeated the same performance, passing the car again, then sitting and looking alert into the distance. This was the only time the male reacted to an auditory stimulus by actually moving towards the source of the sound and not just by looking in its direction. Whilst only a cub reacted to the baboon sound, the male alone reacted to the hyaena call.

It is well known that mammals give an alarm call when noticing a predator, but I have seen no mention in the literature of birds sounding the alarm when sighting lions. However, on several occasions various species of birds were observed to emit calls in situations which made them appear to be alarm signals.

The calls of Marabou storks flying low over a pride on a kill might be interpreted as a signal for other Marabous to assemble for a prospective meal, but the significance of non-scavengers calling when flying over the lions must be sought elsewhere, possibly in an alarm function. On 9th August 1968, 25th March 1969 and 25th April 1969, a pair of crested cranes were seen flying silently, low, towards a pride. When above the lions, they circled a number of times honking loudly, then flew away silently. On one of these occasions they returned five minutes later and repeated this performance. On 15th May 1969 and again on the 21st plovers were seen to behave in a similar manner, whilst on 23rd November 1968 Hadada ibis flew over the lions to settle on a tree 25 m. away, then returned and flew over them again in the opposite direction, calling loudly several times as they passed over the pride. As the birds passed overhead, all the cubs followed their flight with their eyes, and one of them jumped 70 cm. into the air with paw extended, breaking a branch of an *Acacia* tree in the process.

On another occasion crows followed a walking lioness, circling low over her, calling loudly. The lioness occasionally looked up at them but was otherwise unmoved. However, at other times when lionesses were on a kill, they were seen to jump, snap or lash out at low flying birds, especially vultures. High flying vultures, on the other hand, were often observed with interest by both Scarface and the lionesses (see Chapter VII).

Whilst they were occasionally waiting on surrounding trees for the lions to finish their meal, at other times the vultures were nowhere in evidence when a carcase was left by the lions. On 15th May 1969, a partly eaten carcase of a giraffe lay abandoned with no lion in the neighbourhood as far as could be ascertained. Several species of vultures and marabou storks were seen resting on the surrounding trees from 0800 hours until 1130 hours without any of them descending. On other occasions animals that had apparently died, lay for several days before any scavengers started feeding on them. It is possible that the Kenya Meat Commission at nearby Athi River offers ample food to these scavenging birds and thus the food supply in Nairobi Park is of secondary importance.

RESPONSE TO WEATHER

A record was kept of air temperatures to try to establish the tolerance of lions to direct sunlight, i.e. the temperature at which they seek shade. However, the temperatures recorded at the times when they sought cover varied so widely that no clear connection could be established. Whilst they usually seek shade or shelter between 0900 hours and 1000 hours, they occasionally stayed in full sunshine until much later. The cubs, however, always sought shelter or shade before the adult lionesses, possibly because their smaller bodies were quicker to heat up, as a result of the higher ratio of area to mass.

The first response to mild or medium-heavy rain was usually headshaking, followed by retreating into taller grass or among taller vegetation. Only once was a violent reaction to a sudden downpour observed when three lionesses and five cubs scrambled into the undergrowth surrounding a *Balanites* tree, under which they had been resting. The scramble was so sudden and violent, with every animal trying to force its head into the undergrowth, that it resulted in the lions climbing over each others' backs and lying atop one another like the spokes of a wheel surrounding the trunk of the tree. After a few minutes a lioness came out again into the open. The downpour having by then somewhat subsided she sat with her back to the direction of the rain, occasionally shaking her head.

Tigers often rest submerged in water (Schaller, 1966). Whilst lions are said to be good swimmers and not averse to crossing streams (Adamson, 1961), during this study they were never seen to enter water except when playing in shallow rainpuddles, hitting the water and splashing about in it.

IX Discussion and Epilogue

This study was based on daily observations between June, 1968 and 1969 and intermittent observations during the first six months of 1968 and the last six months of 1969. Data from a limited number of night observations are also included. A follow-up study of two years' duration was started in October 1970, as it was felt that valuable additional data may be obtained from this population where every animal was known and could be identified even after long absence from the Park.

Information was sought on the changes that occur within a social unit (pride) and how they come about. Any eventual annual or seasonal cycles in range utilization or predation pattern should also come to light during such a long-range project, as by the end of this second phase a more or less continuous record of five years will be available.

None of the new data so far obtained necessitated any major revisions of conclusions drawn on the evidence presented in this book. Some additional evidence has tended to support previous findings as, for instance, data on additional adult males became available and some new observations confirmed tentative conclusions previously based on circumstantial evidence only as in the case of territoriality.

It appears that a period of major change and subsequent readjustment in population and range occupancy, similar to the one at the beginning of this study, may occur in Nairobi National Park approximately every 2 years. After a relatively stable period a major exodus occurred early in 1970. Towards the second half of that year many of the lions reappeared, but the prides had broken up and the new associations were formed between close blood relations, sister with sister and mother with daughter. All young adult males had separated from their prides and had become either solitary or formed bachelor groups with other young lions of their pride.

This two-yearly readjustment in the population seems to coincide with the length of time it takes for a generation to attain maturity and become independent.

The reproductive history of some of the lionesses between 1968 and 1972 indicates that they normally produce new litters every 2 to $2\frac{1}{2}$ years and that first litters are born to lionesses of about 3 years of age. In case young cubs are lost the mother may mate again soon after loss of the last cub and produce a new litter.

The impression gained during the original study that several litters of cubs are only brought up together if the age difference is no more than about four months, gained additional support during subsequent observations. Age of cubs seems to be one of the mechanisms which determine associations between lionesses in a pride. On several occasions mother and daughter associations broke up when litters of mother and daughter were too far apart in age but were maintained or re-established when both had young at an interval of only two months.

At the beginning of the original study there was some circumstantial evidence to suggest that in the case of lions we may speak of a territory as defined by Burt '. . . part of the home range . . . that is protected from other individuals of the same species.' The fact that the ranges of the two males, Scarface and Kihara, overlapped hardly at all during the period when they were both resident, but that, after the Athi male disappeared, Scarface extended his well over the range of the Athi pride, was one indication of adult males' intolerance of each other. The disappearance from the Park of a former long-time resident male, Spiv, after Scarface established himself in the Park towards the end of 1967, was another. Fresh injuries on the newly arrived male indicated that the change in ownership had resulted from physical encounters rather than merely from threat display and avoidance.

During the years following the 1968–69 investigation there was more definite evidence of territoriality. One encounter that had all the characteristics of a territorial contest, was witnessed in December, 1970. It involved part of the former Romola pride and the Forest pride, which took up residence late in 1969. Both prides consisted of one adult lioness and three young adults of 2 to 3 years. Although it was difficult to follow events as lions chased each other changing directions several times, making it impossible to keep track of individuals, it appears that the Romola pride proved superior as the other group avoided that area from that day onwards and, in fact, restricted its range very severely.

During the 1968–69 study the lions preyed on fifteen species of mammals and two species of birds. The four main species killed were: kongoni, wildebeest, zebra and warthog, together constituting 81% of all kills. Preference ratios calculated on the basis of frequency of selection as against abundance in total prey population showed a decided preference for wildebeest, although of lesser magnitude than shown by Foster and Kearney in 1967. This trend towards a decreasing predation on wildebeest continued during 1970–72 when percentage of wildebeest kills in total kill was down to 12% from 25% in 1968–69. However, the combined percentage of the four main prey species has remained at a surprisingly constant 81% of all kills.

Hirst (1969) suggests two possible causes for high predation on wildebeest evident in many areas where lion predation has been studied: (1) preference for wildebeest as prey, presumably based either on a liking for the flesh of the animal, or on ease of capture, and (2) high incidence of contact brought about by similar habitat preferences. The first cause seems to be supported by data from areas where there is preferential selection of this animal under habitat conditions which differ widely (Pienaar, 1969; Hirst, op. cit.; Foster and Kearney, op. cit.). The latter may apply, in modified form, to the present situation. By day wildebeest are mostly found on the tops of the ridges in open grassy areas and lions in the river valleys; however, at night wildebeest are often in or on the edge of *Acacia* scrub, the preferred hunting ground of lions. It is suggested that the presence of the extensive *Acacia drepanolobium* areas in Nairobi Park contributes to the high predation on the small wildebeest population.

Co-operative hunting has evolved as an adaptation for hunting prey much larger than the predator. Canids (wild dogs, wolves) hunt co-operatively, but have evolved no specialized killing bite. Most felids are solitary hunters and have evolved specialized techniques for killing their prey; the bite in the nape of the neck in some Felinae (Leyhausen, 1956 a) and the stranglehold on the throat used by both cheetah and tiger are such techniques (Schaller, 1967 and 1968). Tigers use both the bite in the back of the neck and the stranglehold on the throat whilst cheetah usually kill by the latter method (Schaller, 1968).

The killing technique of the lion is either strangulation or suffocation, the latter used most frequently for species larger than the predator (Schaller, pers. com.; Guggisberg, 1961). Of these two, suffocation—clamping the mouth over the victim's muzzle—is the more specialized technique and has not been described for any other felid.

Both tiger and cheetah usually hunt alone whilst lions often hunt co-operatively. It is suggested that the suffocating muzzle-hold as a killing method has evolved during co-operative hunting as instances of a predator holding on to the victim's nose have only been described for the lion and the wolf, both hunting in groups, although in the latter case not as a killing, but only as a distracting device. This suffocating killing method is extremely efficient and has the advantage of greater speed—whilst retaining all the advantages of the stranglehold on the throat—thus reducing the effort required and the probability of injury through the victim's struggles. The proposition is put forward that this method may evolve into 'the killing technique' of the lion—as the bite in the nape is 'the killing bite' for some Felinae (Leyhausen, 1956 a)—and so reduce the survival value of social hunting. As observations during this study have shown, by this method a single animal can deal quickly and efficiently even with prey, such as wildebeest, much larger than the predator.

Although predator populations are limited by the prey as well as by social factors (Hairston et al. 1960) the limiting factors are not so much the numbers of prey animals but their vulnerability. (Schaller, 1967; Wright, 1960). Thus fewer prey in broken terrain may support as many predators as greater numbers in open plains. The great variety of terrain in Nairobi Park, its broken character and the large area of *Acacia drepanolobium* scrub, favourable to the lions' hunting technique, can support a relatively high density of lions compared to other areas (see Chapter VII).

The largest number of resident lions during the last 20 years was reported to be thirty-five by S. Ellis, former Warden of the Park. Shortly after the termination of this study in 1969, the lion population suddenly increased to over forty by (1) births, (2) return of some previous residents, (3) immigration of new prides. However, this situation was short-lived as a number of lions left the Park within a few months and the total is now (1972) thirty-six, with 66% of the population under 2 years of age. This shows a consistently high percentage of juveniles in this Park. Cub mortality during 1968–69 was estimated at 15%. This represents a high breeding success when compared with zoo data where mortality may be as high as 70%. Even under good conditions the juvenile mortality rate in the wild has been estimated by Guggisberg (1961) as 50% although this figure includes mortality up to 2 years and it is believed to be quite high at about 2 years when males usually leave their prides. As the oldest cubs in this study were only 15 months old by the end of regular observations and left the Park when just under 2 years of age, no complete juvenile mortality data could be calculated. It is, moreover, possible that mortality of newborn cubs (under two months and before they were first seen) would raise this figure somewhat.

Two possible reasons for the present seemingly very low cub mortality are suggested: (1) As the Park, with access only from the south, is virtually a cul-de-sac, few, if any, transient males pass through: none were seen during this study. In contrast, in areas such as the Serengeti Plains, such males may constitute a hazard to small cubs (Schaller, 1969). (2) The scarcity of hyaenas in the Park may also lessen the danger to survival of cubs. Hyaenas are known to kill and devour aged and infirm lions (Cullen, 1969) and, in areas where they are numerous, probably also kill cubs.

The present study indicates a considerable interchange between the Park population and the lions outside. Immigration and emigration seem to reach a peak every 2 years, but even during their time of stable residence, the prides make considerable use of the Kitengela area south of the Park. On several occasions Park lions were actually seen in those areas and on many more footprints indicated that they had crossed the Embakasi river.

If this interchange were made impossible either by fencing or by close settlement along the southern boundary, it would probably affect the lion population in several ways. (1) through possible changes in the prey population, (2) by prohibiting dispersal of surplus individuals, (3) by restricting gene flow.

(1) At present there is a considerable movement of herbivores out of the Park during the wet season and back into the Park during the dry periods. The yearly fluctuations of the herbivore population range from a low of about 3,000 during the April-May 'long rains' to about twice that number by the end of the long dry season in October. It is hazardous to try to predict what effect restriction to the Park would have on the herbivores. The present migratory system amounts to a natural grazing rotation. It is quite possible that this is essential to keep the Park at its optimum carrying capacity and that stopping such a rotation system would lead to over-grazing and a consequent lowering of the area's carrying capacity. A decline in the herbivore population may, in turn, affect the status of the predators.

(2) Even from a large Park like the Kruger National Park in South Africa, there is a constant emigration into the surrounding areas even though lions are hunted outside the Park (Pienaar, 1969). This need for a dispersal area for the surplus population would be even more acute in a small unit such as Nairobi Park. As both Wynn-Edwards (1964) and Hairston et al. (1960) recognize, predator populations, although generally food limited, are, in certain species, also limited by social factors. Wynn-Edwards (op. cit.) lists some of the means by which such a control is effected as (1) territoriality, and (2) competition for membership in a group whereby only a certain number is accepted, the surplus must move away. In lions both of these mechanisms seem to operate, the second, as in most species, applies mostly to young males who are forced out of the pride and often have to emigrate to other areas.

If emigration became impossible, and the natural increase would have no possibility of disposal, other social mechanisms to control the population would come into play, such as increased intolerance and reduced reproductive rate and success (Schaller, 1967). The present high percentage of juveniles could not then be maintained and as lion cubs are a great attraction to the visiting public, the overall allure of the Park would diminish.

(3) If no interchange with lions outside the Park were possible, inbreeding in the Park population might lead to the spread of such undesirable characteristics at least (from the Park's point of view) as manelessness exemplified by Kihara and several of his offspring.

As far as the lions are concerned it therefore seems highly desirable that the area to the south of the Nairobi Park should be preserved in something like its present form. If this does not occur the famed Nairobi lions are likely to decline in both numbers and vigour.

Appendices

Vegetation types related to soils as classified by Verdcourt (in Heriz-Smith, 1962).

(1) Black cotton soil (margalithic earth), neutral or alkaline, associated with *Acacia drepanolobium*. The trees are mostly dwarfed, occurring in varying density associated with the following grasses: *Pennisetum mezianum, Bothriochloa insculpta, Themeda triandra* and *Digitaria macroblephara*.

(2) Reddish loam with grassland and scattered trees. Main trees on this association found on areas of the park between the south boundary and the black cotton soil are: *Acacia gerrardii var. latisiliqua, Acacia seyal, Acacia mellifera, Acacia tortilis, Commiphora trothae, Albizia isenbergiana, Balanites aegyptiaca, Combretum gueinzii, Ziziphus mucronata*. The mature grasslands are characterized by *Themeda triandra, Bothriochloa insculpta, Digitaria macroblephara,* and *Pennisetum mezianum*.

(3) Forest and woodland in the higher areas of the Park on red soil with open areas of grasses and scattered shrubs. Main tree species of the forest are: *Croton megalocarpus, Olea africana, Schrebera alata, Brachylaena hutchinsii*.

(4) Acacia groves with *Acacia kirkii* occurring in the drier valleys.

(5) Riverine forest where *Acacia xantophloea* is predominant. Thickets composed of *Phyllanthus sepialis, Carissa edulis* and *Croton dichogamus* are often associated with these trees.

(6) Bushland and thicket is found on shallow soils on sides of dry stream beds and gorges. Main species recorded here are: *Grewia similis, Lippia javanica, Ziziphus abyssinica, Lantana viburnoides, Cordia ovalis,* with coarse grasses such as *Themeda* and *Cymbopogon* underneath the shrubs.

On the mounds formed by old termitaria the most characteristic association is *Achyrantes aspera* and *Pavonia patens*.

APPENDIX II

List of some of the most important mammals in NNP

Herbivores			
Bushbuck	Hippo	Steinbuck	Waterbuck
Cheetah	Hyaena	Thomson's gazelle	Wildebeest
Dik-dik	Impala	Warthog	Zebra
Eland	Jackal		
Giraffe	Kongoni (Hartebeest, Coke's)	*Carnivores*	*Primates*
Grant's gazelle	Reedbuck	Leopard	Baboon (Olive)
	Rhino	Lion	Vervet Monkey

APPENDIX III

Identification method

Vibrissae on mammals were grouped into five categories by Pocock (1914): (1) buccal, (2) interramal, (3) genal, (4) superciliary, (5) subocular. In the Felidae all of these, except the interramal on the chin, are present. The buccal vibrissae consist of : (a) those on the muzzle and upper lip, the mysticials, and (b) those on the chin and lower lip, the submentals. The mysticials are usually arranged in definite longitudinal lines and in felids are accompanied by rows of dark spots. The number and arrangement of the vibrissae do not match exactly the number and arrangement of the spots.

In lions there are about six longitudinal rows of spots arranged at fairly regular intervals parallel to the upper lip. Only the upper four or five rows are clearly discernible, as those close to the upper lip become indistinct. Rows 'b' to 'd' (Fig. A–1) consist of between six to twelve spots each but the exact number is not easy to establish as row 'b' and 'd' curve towards each other at the caudal end, forming a semicircle. Thus it is difficult to say where one row ends and the next begins.

Row 'a' is considerably shorter consisting of a maximum of five spots, or, in some cases, none at all (five was the maximum number seen in eighty-six sides). These are spaced at irregular intervals and the two sides are not symmetrical. The number and location of these uppermost spots in relation to row 'b' form the basis of identification.

Vibrissae emerge only from a limited number of these mysticial spots, mostly from those on the caudal end of rows 'b' to 'd'; those in row 'a' were only once seen to mark the emergence of a vibrissa.

Method of application of recognition system

Pencil sketches were made, in the field, of the two upper rows of spots onto a schematic outline of a lion's profile from both sides. Other identifying marks, if any, such as torn ears and faulty dentition, were also noted on the full face outline. The animal was given a number and name. The approximate age, the sex and the pride to which it belonged, were also noted.

Black and white photographs were taken from both sides of the face as close to a 90 degree angle as possible. These photographs were then enlarged and checked against the pencil sketches. The pattern was determined by deciding whether a spot in row 'a' lay directly above a spot or between two spots in row 'b'. This was as fine a distinction as to position as could be made with any degree of certainty and was found to be satisfactory. As an aid to this decision an approximately 2 mm. gauge wire mesh in a jointed frame was set over the photograph to help line up rows 'a' and 'b'. On a schematized plan (Fig. A–2) positions were numbered beginning at the buccal end. Each number in row 'b' marks the position of a spot in that row. This serves as a baseline for positioning spots in row 'a', where nineteen possible positions are marked.

Although row 'b' is, as a rule, quite regular, with spots evenly spaced and the line running fairly straight except for the slightly downward curving first three spots, there are occasional irregularities in spacing or direction (Plate 2 c and d). In these cases the spots were placed, on the schematized drawing, below or above the midline, where they would usually be entered. The same applies to row 'a' where such irregularities in spacing and deviations from the straight line are much more common. Occasionally an extra spot occurs between two rows, usually between lines 'b' and 'c'.

Fig. A–1

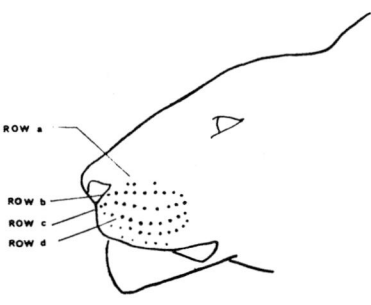

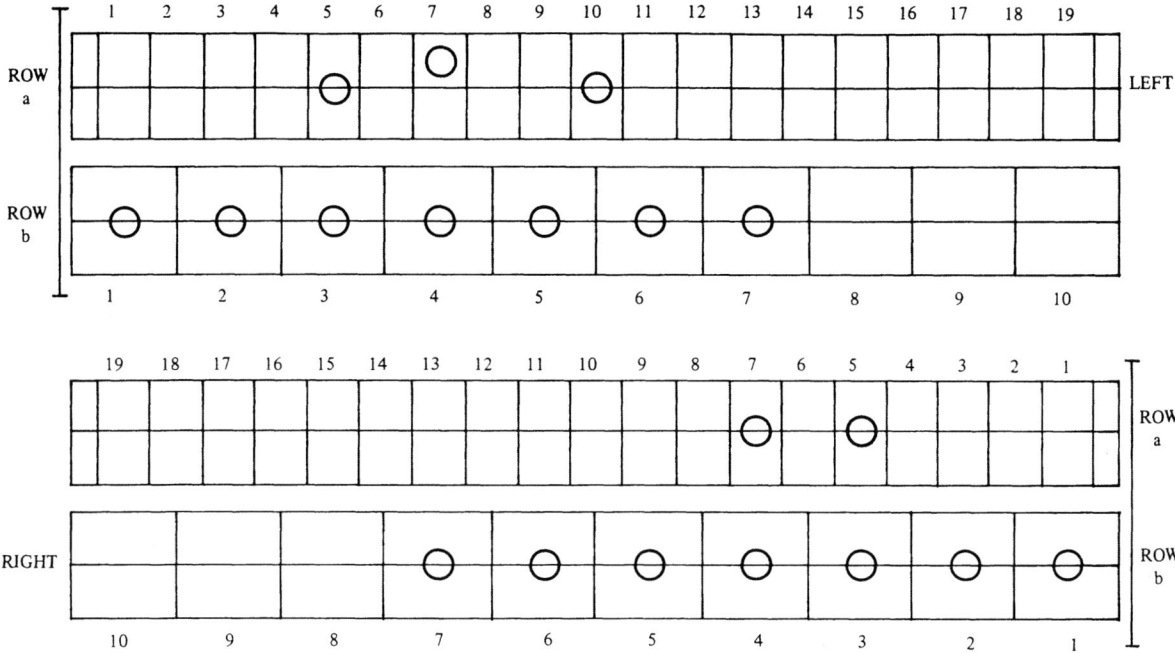

Fig. A-2

Records on some of these animals have now been kept for four years (January 1968, to December 1971) and no change in pattern has been observed, except for the one case mentioned below. Photographs in Plate 2 a and b were taken 27 months apart and show no sign of change, although the first photograph was taken of an immature animal, thus increasing the chances of change. The only case of an apparent alteration so far observed was the appearance, on a 9-month-old cub, of a spot not previously observed. When rechecking the original photographs more carefully, taken at the age of 4 months, it became apparent that the spot was present but too faint to be identifiable as a vibrissal spot at that time. This was the only such instance recorded and seems to be due to developmental growth rather than genuine change in pattern. This leads me to assume that the pattern, once established, can be considered permanent.

Pennycuick and Rudnai (1970) have calculated the reliability of the method. At the 1% level all but two of twenty-five animals in this study could be reliably identified by the spot pattern alone. In two individuals the chances of making a mistake would be somewhat higher than 1%, for these additional data, such as sex of animal and scars or nicks in the ears would have to be used as supplementary bits of information to arrive at the arbitrarily set criterion of reliability at the 1% level.

APPENDIX IV

Previous studies

Studies based mostly on observations of captive animals are: J. Krumbiegel's 'Der Löwe', (1952), a general treatment of the lion; J. Cooper's 'An exploratory study on African lions' (1942), giving an account on fighting, sexual and feeding behaviour of captive lions; R.M.F.S. Sadleir's paper (1966) on the reproduction in the larger Felidae in captivity.

Glickman (1964) discusses the development of curiosity in the genus *Panthera* on the basis of experiments conducted in zoos, and D. D. Davis (1962) compares anatomical development in cats and lions.

Another treatment of anatomical characteristics is Weigel's paper (1961) on the colour and pattern of the felids, mentioning a unique case of a lioness (Schneider, 1949) which, after having raised a litter of cubs and while still coming into oestrus, grew a distinct mane.

Leyhausen (1956) discusses different behaviour patterns in the various genera and species of felids and describes the anatomical and behavioural differences in hybrids of tigers and lions (1950).

Studies on wild populations include Grzimek's (1960) experiments with lions and zebra dummies, Kühme's (1966) observations on the social habits of lions, both in the Serengeti Park; Makacha and Schaller's (1969) paper on lions in Lake Manyara Park and R. Schenkel's (1966) discussion on territoriality and marking in lions and rhinos.

Cowie's (1966) book 'The African lion' is a generalized description of the wild lion and its habits, while Adamson (1960, 1961, 1962) and Carr (1962) describe the release of tame animals.

Studies on predation including data on lions were published by Wright (1960); Foster and Kearney (1967); Kruuk and Turner (1967); Pienaar (1969); Mitchell et al. (1965); Foster and Coe (1968) and Hirst (1965).

APPENDIX V

Taxonomic standing

Order: *Carnivora* (Bowdich)

Superfamily: *Feloidea* (Simpson)

Family: *Felidae* (Gray)

Subfamily: *Felinae* (Trouessart)
Small, medium to large forms; completely retractile claws with protective skin folds. Ossified hyoid, slitlike pupils.

Subfamily: *Lyncinae* (Gray)
Medium sized cats with short tail, retractile claws with protective skin fold, ossified hyoid, tuft of hair at tip of pointed ears.

Subfamily: *Acinonychinae* (Pocock)
Large cats with non-retractile claws, no protective skin fold, ossified hyoid.

Subfamily: *Pantherinae* (Pocock)
Large cats, completely retractile claws, well developed protective skin fold, hyoid incompletely ossified, oval pupils.

    Genus: *Panthera* (Oken)
                 Panthera leo (Linnaeus)
                 Panthera tigris (Linnaeus
                 Panthera onca (Linnaeus)
                 Panthera pardus (Linnaeus)

    Genus: *Uncia* (Gray)
                 Uncia uncia (Schreber)

Classification according to Haltenorth (1953)

List of proposed subspecies considered valid by Guggisberg (1961).

Indian lion: *Leo leo goojratensis* (Smee)
Barbary lion: *Leo leo leo* (Linnaeus)
Senegal lion: *Leo leo senegalensis* (Meyer)
Masai lion: *Leo leo massaicus* (Neumann)
Kruger lion: *Leo leo krugeri* (Roberts)
Kalahari lion: *Leo leo vernayi* (Roberts)
Cape lion: *Leo leo melanochaitus* (Hem. Smith) or *Leo leo capensis* (Fischer)

APPENDIX VI

Meteorological data

The following meteorological data have been kindly supplied by the East African Meteorological Department: the figures are from the Embakasi weather station located just north-east of Nairobi National Park.

The annual mean temperature for the year of the study, June 1968 to July 1969, was 18·4°C. Fig A–3 shows the slight fluctuation of the monthly mean temperatures throughout the year.

Rainfall in East Africa is seasonal, and in the Nairobi area is divided into two rainy periods: the 'long' rains from March to May and the 'short' rains from October to December. Table A–2 shows the monthly rainfall for the study period and monthly averages for the past twelve years. The monthly mean was 63·8 mm. for the study year as compared to 68·8 mm. for the past twelve years.

Monthly means for hour/day sunshine ranged from a low of 2·8 hours for July and August to a high of 9·0 and 9·5 for December and January, respectively.

Table A–1

	Monthly rainfall mm	Average for 12 years mm
1968—June	68	24·5
July	1·9	6·4
August	1·9	16·7
September	—	14·3
October	34·5	48·4
November	278·6	164·9
December	54·6	91·6
1969—January	69·7	50·2
February	59·9	50·9
March	38·0	80·8
April	8·7	151·7
May	150·0	125·0
Monthly mean:	63·8	68·8

Figures from Embakasi meteorological station.

Fig. A–3

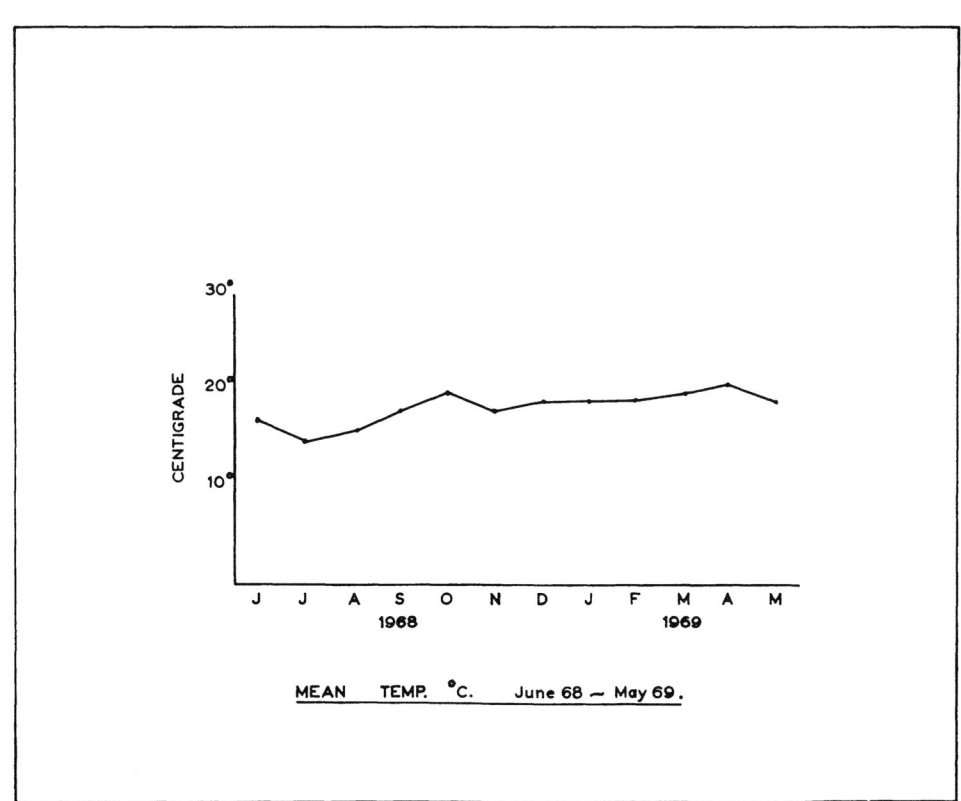

MEAN TEMP. °C. June 68 – May 69.

APPENDIX VII

Glossary of common names

Baboon	*Papio anubis* (Matschie)
Bat-eared fox	*Otocyon megalotis* (Miller)
Bear, brown	*Ursus arctos* (Linnaeus)
grizzly	*U. horribilis* (Merriam)
polar	*Thalarctos maritimus* (Phipps)
Buffalo	*Syncerus caffer* (Sparmann)
Bushbuck	*Tragelaphus scriptus* (Pocock)
Caribou	*Rangifer arcticus* (Hollister)
Cat, domestic	*Felis catus* (Linnaeus)
Cheetah	*Acynonix jubatus* (Heller)
Cow, domestic (cattle)	*Bos taurus* (Linnaeus)
Chimpanzee	*Pan troglodytes* (Giglioli)
Crane, crested	*Balearica regulorum* (Reichenow)
Crocodile	*Crocodylus niloticus* (Laurenti)
Crow, pied	*Corvus albus* (Müller)
Dik-dik	*Rhynchotragus kirkii* (Thomas)
Dog, wild	*Lycaon pictus* (Thomas)
domestic	*Canis familiaris* (Linnaeus)
Eagles	*Falconides*
Eland	*Taurotragus oryx* (Lydekker)
Elk (Amer.)	*Cervus canadensis* (Erxleben)
Fox, bat-eared	*Otocyon megalotis* (Miller)
Gazelle, Grant's	*Gazella granti* (Heller)
Thomson's	*G. thomsoni* (Gunther)
Genet	*Genetta genetta* (Matschie)
Giraffe	*Giraffa camelopardalis* (Matschie)
Gull, laughing	*Larus atricilla* (Linnaeus)
Hare	*Lepus capensis* (De Winton)
Hartebeest, see kongoni	
Hippo	*Hippopotamus amphibius* (Linnaeus)
Horse, domestic	*Equus caballus* (Linnaeus)
Hyaena (spotted)	*Crocuta crocuta* (Matschie)
Ibis, Hadada	*Hagedashia hagedash* (Latham)
Impala	*Aepyceros melampus* (Matschie)
Jackal (silverback)	*Canis mesomelas* (Heller)
(side-striped)	*Canis adustus* (Sundevall)
Jaguar	*Panthera onca* (Linnaeus)
Kongoni (hartebeest)	*Alcephalus buselaphus* (Thomas)
Kudu (greater)	*Tragelaphus strepsiceros* (Pallas)
(lesser)	*Strepsiceros imberbis* (Blyth)
Leopard	*Panthera pardus* (Linnaeus)
Lion	*Panthera leo massaica* (Neumann)
Meerkat	*Suricata suricatta* (Schreber)
Monkey, vervet	*Cercopithecus aethiops* (Pocock)
Moose	*Alces alces* (Gray)
Ostrich	*Struthio camelus* (Linnaeus)
Pig, domestic	*Sus scrofa* (Linnaeus)
Plover, crowned	*Stephanibyx coronatus* (Boddaert)
Puku	*Adenota vardoni* (Sclater)
Puma	*Puma concolor* (Linnaeus)
Reedbuck	*Redunca fulvorufola* (Rothschild)
Rhinoceros	*Diceros bicornis* (Linnaeus)
Spring hare	*Pedetes surdaster* (Hollister)
Steinbuck	*Raphicerus campestris* (Matschie)
Stork, marabou	*Leptoptilos crumeniferus* (Lesson)
Tiger	*Panthera tigris* (Linnaeus)
Tick	Sub-order: *Ixodidae, Argosidae*
	Class: *Arachnidae*

Vultures	*Aegypiinae*
Waterbuck	*Kobus defassa* (Heller)
Warthog	*Phacocoerus aethiopicus* (Cretzshmar)
Wildebeest	*Connochaetes taurinus* (Thomas)
Wolf	*Canis lupus* (Goldman)
Zebra	*Equus burchelli* (Lydekker)

References

ABLES, E. D. (1969) Home-range studies of red foxes *(Vulpes vulpes)*. *J. Mammal* 50 (1):108–120

ADAMSON, J. (1960) *Born Free*. Pantheon Books, New York

ADAMSON, J. (1961) *Living Free*. Harcourt, Brace & World, New York

ADAMSON, J. (1962) *Forever Free*. Collins & Harvill Press, London

ARMSTRONG, E. W. (1950) The nature and function of displacement activities. *Symp. Soc. Exp. Biol.* 4:361

BAERENDS G. P. (1956) Aufbau des tierischen Verhaltens. *Handbuch der Zoologie*, Walter de Gruyter & Co., Berlin

BEATON, K. de P. (1949) *A warden's diary*. East African Standard Ltd., Nairobi

BEN SHAUL, MILLER, D. (1962) The composition of the milk of wild animals. *Int. Zoo Yb.* 4:333

BOURLIERE, F. (1965) Densities and Biomass of some Ungulates in E. Congo and Rwanda with notes on population structure and lion/ungulate ratios. *Zoologica Africana I* (1):199–207

BRAIN, C. K. (1965) Observations on the behaviour of vervet monkeys *(Cercopithecus aethiops)*. *Zoologica Africana I* (1):13–27

BROMLEY, P. T. (1969) Territoriality in pronghorn bucks on the National Bison Range, Moiese, Montana. *J. Mammal.* 50 (1):81–89

BURT, W. H. (1943) Territoriality and home range concepts as applied to mammals. *J. Mammal.* 24:346–352

CARR, N. (1962) *Return to the Wild*. E. F. Dutton & Co., New York

COLE, D. D., *et al.* (1966) A study in social dominance in cats. *Behaviour* 27:39–53

COOPER, J. (1942) An exploratory study on African lions. *Comp. Psychol. Monograph* 17 (7):1–48

COWIE, M. (1951) *The Nairobi Royal National Park Guidebook*. The Trustees of the Royal National Parks of Kenya

COWIE, M. (1966) *The African Lion*. The Golden Press, New York

COWAN, I. M. (1947) The timber wolf in the Rocky Mountain National Parks of Canada. *Can. J. Res.* 25:139–174

CRISLER, L. (1956) Observations of wolves hunting Caribou *(Rangifer arcticus)*. *J. Mammal.* 37:337–346

CULLEN, A. (1969) *Window into wilderness*. East African Publishing House

DASSMAN, R. and MOSSMAN, A. (1962) Population studies of Impala in S. Rhodesia. *J. Mammal.* 43 (3):375

DAVID, R. (1962) Notes on hand rearing various species of mammals. *Int. Zoo Yb.* 4:300

DAVIS, D. D. (1962) Allometric relationship in lions vs. domestic cats. *Evolution* 16:505–514

DOWSETT, R. J. (1966) Wet season game population and biomass in the Ngoma area of the Kafue National Park. *The Puku* 4

ELOFF, F. C. (1964) On the predatory habit of lions and hyaenas. *Koedoe* 7:105–112

ERRINGTON, P. (1946) Predation and vertebrate populations. *Quart. Rev. of Biol.* 21 (2):144–177

ESTES, R. D. (1967) Predators and scavengers. *Nat. Hist.* 76 (2):20–37

ESTES, R. D. and GODDARD, I. (1967) Prey selection and hunting behaviour of the African wild dog. *J. Wildl. Mgmt.* 31 (1):52–70

EWER, R. F. (1959) Suckling behaviour in kittens. *Behaviour* 15:1–2

EWER, R. F. (1968) *Ethology of Mammals*. Logos Press, Ltd., London

FIEDLER, W. (1957) Beobachtungen zum Markierungverhalten einiger Saugetiere. *Z. Saugetierk.* 22:57–76

FORBES, R. B. (1963) Care and early behaviour development of a lion cub. *J. Mammal.* 44:110–111

FOSBROOKE, H. (1963) The stomoxys plague in Ngorongoro, 1962. *E. Afr. Wildl. J.* 1:124–126

FOSTER and COE (1968) Biomass of game animals in Nairobi National Park. *J. of Zool.* 155: 413–425

FOSTER and KEARNEY (1967) Nairobi National Park game census, 1966. *E. Afr. Wildl. J.* 5:112–120

GLICKMAN, S. E. (1964) The development of curiosity within the genus *Panthera*. *Zool. N.Y.* 49 (2):109–114

GOODALL (VAN LAWICK), J. (1967) Mother-offspring relationships in free-ranging Chimpanzees. *Primate Ethology*, ed. D. Morris. Weidenfeld & Nicholson, London

GRZIMEK, B. (1960) Attrappenversuche mit Zebras und Löwen in der Serengeti. *Z. Tierpsychol.* 17:351–357

GUGGISBERG, C. A. W. (1961) *Simba*. Howard Timmins, Cape Town

HAIRSTON, N. G. SMITH and SLOBODKIN (1960) Community structure, population control and competition. *Amer. Nat.* 94:421

HALTENORTH, T. (1953) *Wildkatzen*. Leipzig

HERIZ-SMITH, S. (1962) *The wild flowers of the Nairobi Royal National Park*. D. A. Hawkins Ltd., Nairobi

HIRST, S. M. (1965) Ecological aspects of big game predation. *Fauna and Flora* 16:3–14

HIRST, S. M. (1969) Population in a Transvaal Lowveld Nature Reserve. *Zoologica Africana* 4 (2):199–230

HOOFF, J. A. R. A. M. VAN, (1965) A large litter of lion cubs. *Int. Zoo Yb.* 5:116

JENSEN, et al. (1968) Sex differences in development of independence in infant monkeys. *Behaviour* 30 (1):1

JORDAN, P. A. et al. (1967) Numbers, turnover and social structure of the Isle Royale wolf population. *Am. Zoologist* 7:233–253

KLEIMAN, D. (1966) Scent marking in Canidae. *Play, exploration and territoriality in mammals*, ed. P. A. Jewell and C. Loizios. Academic Press, London

KLINGEL, H. (1967) Sozial Organisation und Verhalten freilebender Steppenzebras. *Z. Tierpsychol.* 24 (5):580–624

KLINGEL, H. and V. (1966) Tooth development and age determination in the plains zebra. *Der Zool. Garten.* 33 (1/3):34–35

KRUMBIEGEL, I. (1952) *Der Löwe*. Neue Brehmbücherei, Leipzig

KRUUK, H. (1966) Clan system and feeding habits of spotted hyaena. *Nature* 209: 1257–1258

KRUUK, H. (1972) *The Spotted Hyaena*. University of Chicago Press.

KRUUK, H. and TURNER, M. (1967) Comparative notes on predation by lion, leopard, cheetah and wild dog in the Serengeti area, East Africa. *Mammalia* 31 (1):1–27

KÜHME, W. (1964) Über die soziale Bindung innerhalb eines Hyänenhundrudels. *Die Naturwissenschaften* 23:567

KÜHME, W. (1964) Freilandbeobachtungen an Löwen und Hyänenhunden in Serengeti National Park, Tang. *Freunde des Kölner Zoo* 7:106–108

KÜHME, W. (1965 a) Communal food distribution and division of labor in hunting dogs. *Nature* 205:443–444

KÜHME, W. (1965 b) Freilandstudien zur Soziologie des Hyänenhundes. *Z. Tierpsychol.* 22 (5):495–541

KÜHME, W. (1966) Beobachtungen zur Soziologie des Löwen in der Serengeti Steppe. *Z. Säugetierk.* 31 (3):205–213

LABUSCHAGNE, R. J. and VAN DER MERWE, N. J. (1963) *Mammals of the Kruger and Other National Parks*. National Parks Board of Trustees of the Republic of South Africa

LE BOEF B. J. (1967) Interindividual association in dogs. *Behaviour* 29 (2-4):218

LEYHAUSEN, P. (1950) Beobachtungen an Löwen-Tiger Bastarden mit einigen Bemerkungen zur Systematik der Grosskatzen *Z. Tierpsychol* 7 (1):48–83

LEYHAUSEN, P. (1956 a) Das Verhalten der Katzen. *Handbuch der Zoologie* 8 (10):1–34

LEYHAUSEN, P. (1956 b) Über die unterschiedliche Entwicklung einiger Verhaltens- weisen bei den Feliden. *Saugetierk. Mitt.* 4 (3):123–125

LEYHAUSEN, P. (1965 a) The communal organization of solitary mammals. *Symp. Zool. Soc. Lond.* 14:249–494

LEYHAUSEN, P. (1965 b) Über die Funktion der relativen Stimmungshierarchie. *Z. Tierpsychol.* 22 (4):412–494

LEYHAUSEN, P. (1965 c) Breeding the black-footed cat, *F.nigripes* in captivity. *Int. Zoo Yb.* 5:178

LINN, D. (1965) Some perspectives in mammal ecology. *Zoologica Africana* I (1):181–192

LORENZ, K. (1963) *On Aggression.* Methuen & Co. Ltd., London

MABERLY, C. T. A. (1965) *Animals of East Africa.* Howard Timmins, Cape Town

MAKACHA, S. and SCHALLER, G. B. (1969) Observations on lions in the Lake Manyara National Park, Tanzania. *E. Afr. Wildl. J.* 7:99–103

MAZAK, V. (1964) A note on the lion's mane. *Z. Säugetierk.* 29 (2):124–126

MAZAK, V. (1965) *Der Tiger.* Kosmos-Verlag, Die neue Brehm-Bücherei, Stuttgart

MECH, L. D. (1966) The wolves of Isle Royale. *Fauna of the National Parks of the U.S. Fauna Series* 7

MEINERTZHAGEN, R. (1938) Some weights and measurements of large mammals. *Proc. Zool. Soc. Lond. Ser. A* 108:433–439

MICHAEL, R. P. (1961) Observations upon the sexual behaviour of the domestic cat *(Felis catus)* under laboratory conditions. *Behaviour* 18 (1/2):1–24

MITCHELL, B. L. *et al.* (1965) Predation on large mammals in Kafue National Park, *Zambia. Zoologica Africana* I (2): 297–318

MURIE, A. (1944) The wolves of Mount McKinley. *U.S. National Park Service Fauna Ser. 5,* 238 pp.

PALEN, G. and GODDARD, J. (1966) Catnip behaviour and oestrus in the cat. *Anim. Behav.* 14:372

PENNYCUICK, C. and RUDNAI, J. (1970). A method of identifying individual lions, *Panthera leo,* with an analysis of the reliability of identification. *J. Zool. Lond.* 160: 497–508

PIENAAR, DE V. (1968) The ecological significance of roads in a National Park. *Koedoe* 11:169

PIENAAR, DE V. (1969) Predator-prey relationship amongst the larger mammals of the Kruger National Park. *Koedoe* 12:108–176

POCOCK, R. I. (1914) On the facial vibrissae of mammalia. *Proc. Zool. Soc. Lond.* 60:889–912

POCOCK, R. (1917) The classification of existing Felidae. *Am. Mag. Nat. Hist.* 20: 329–350

PRECHT, H. (1952) Jumping spiders. *Z. Tierpsychol.* 9:207

RABB, G. B. *et al.* (1962) Comparative studies of canid behaviour IV. Mating behaviour in relation to social structure in wolves. *Amer. Zool.* 2 (3):440

ROBINETTE, W. L. (1961) Notes on cougar productivity and life history. *J. Mammal.* 42 (2):204–217

ROTH, H. H. (1965) Observations on growth and ageing of warthog. *Z. Säugetierk* 30 (46):164

ROWLAND WARD, (1962, 1966) *Records of big game, Africa,* London, 374 pages

SACHS, R. and DEBBIE, J. G. (1969) A field guide to the recognition of parasitic infesta- tions in game animals. *E. Afr. Wildl. J.* 7:27–37

SADLEIR, R. M. F. S. (1966) Notes on reproduction of the larger Felidae. *Int. Zoo Yb.* 6:184–193

SCHALLER, G. B. (1966) The tiger and its prey. *Nat. Hist.* 75 (8):30–37

SCHALLER, G. B. (1967) *The deer and the tiger.* The University of Chicago Press, Chicago and London

SCHALLER, G. B. (1968) Hunting behaviour of the cheetah in the Serengeti National Park, Tanzania. *E. Afr. Wildl. J.* 6:95–100

SCHALLER, G. B. (1969) Life with the king of beasts. *Natn. Geogr.* 135 (4):494–519

SCHALLER, G. B. and LOWTHER, G. (1969) The relevance of carnivore behaviour to the studies of early hominids. *Anthropol.* 25:307–341

SCHENKEL, R. (1966) Play, exploration and territoriality in the wild lion. *Symp. Zool. Soc. Lond.* 18:11–12

SCHENKEL, R. (1966) Zum Problem der Territorialitat und des Markierens bei Saugern am Beispiel des Schwarzen Nashorns und des Löwen. *Z. Tierpsychol.* 23 (5)

SCHEUNERT-TRAUTMAN, A. (1957) *Lehrbuch der Veterinär Physiologie.* Paul Parey, Hamburg

SCHLOETH, R. (1961) Das Sozialleben des Camargue-Rindes. *Z. Tierpsychol.* 18 (5): 590–593

SCHNEIDER, K. M. (1932) Flehmen III. *Der Zool. Garten* N F 5:220–226 (quoted by Leyhausen)

STENLUND, M. H. (1955) A field study of the timber wolf *Canis lupus* on the Superior National Forest, Minnesota. *Minn. Department of Conservation, Technical Bulletin No. 4*

STEVENSON-HAMILTON, J. (1937) *South African Eden; From Saabi Game Reserve to Kruger National Park.* London

STEWART, D. R. M. and J. (1963) The distribution of some large mammals in Kenya. *J. E. Afr. Nat. Hist. Soc.* 24 (3) (107)

STRUHSAKER, T. T. (1967) Social structure among vervet monkeys. *Behaviour* 29 (2-4): 83–121

URSIN, H. (1964) Flight and defence behaviour in cats. *J. Comp. Physiol.* 58 (2): 180–186

VAN VLECK, D. (1969) Standardization of *Microtus* home range calculations. *J. Mammal.* 50 (1):69–80

WALKER, E. P. (1968) *Mammals of the World.* The Johns Hopkins Press, Baltimore

WALTHER, F. R. (1963) Verhaltensstudien an der Gattung Tragelaphinae. *Z. Tierpsychol.* 21 (4)

WALTHER, F. R. (1969) Verhaltensstudien an der Grantgazelle in Ngorongoro Krater. *Z. Tierpsychol.* 22:167

WEIGEL, VON I. (1961) Färbung und Musterung der Felidae. *Säugetierk. Mitt.* 9:74–78

WEIR, J. and DAVISON, E. (1965) Daily occurrence of African game animals at water holes during dry weather. *Zoologica Africana* 1 (2):353–367

WILLIAMS, JOHN G. (1967) *A field guide to the National Parks of East Africa.* Collins, London

WRIGHT, B. S. (1960) Predation on big game in East Africa. *J. Wildl. Mgmt.* 24:1–15

WYNN-EDWARDS, V.C. (1964) Population control in animals. *Scient. Amer.* 211 (2):68–76 Boyd, Edinburgh and London

WYNN-EDWARDS, V.C. (1964) Population control in animals. *Scient. Amer.* 211 (2):68–76

YOUNG, E. (1966) Nutrition of wild S. African felines and some viverrids. *Afr. Wildlife* 20 (4)

YOUNG, S. and GOLDMAN, E. (1946) *The puma, mysterious American cat.* Dover Publications, New York

ZUCKERMAN, S. (1932) *The social life of monkeys and apes.* Kegan Paul, Trench, Trubner & Co. Ltd., London

ZUCKERMAN, C. B. (1953) The breeding seasons of mammals in captivity. *Proc. Zool. Soc. Lond.* 122 (4):827–947

Acknowledgements

This study was made possible by a grant from the Ford Foundation made available to me through the Zoology Department of University College, Nairobi. I want to express my appreciation for this assistance, as well as for the permission granted by the Director of the Kenya National Parks, Mr. Perez Olindo, to conduct research in Nairobi Park. I also want to express my gratitude to the Wardens and Management of Nairobi National Park for their assistance and cooperation.

My sincere thanks go to the many friends who helped me during this study in so many ways: by being night drivers, which was not always an easy task, and by giving advice, assistance and encouragement. I would like to list their names in alphabetical order: L. H. Bonnett, Ian and Lise Campbell, Michael Fennessy, Giovanna Hoffman, Bill Hurst, T. D. Morris, Valentina Rosselli, Peter Squelch, John Wyatt.

Dr. Malcolm Coe and Dr. John Sale gave encouragement and advice in the initial stages of the study and Dr. C. Pennycuick gave invaluable help in guiding the project through its most difficult periods. Morris Gosling, Dr. Hans Klingel and Dr. Murray Watson's help with sexing and ageing of the prey animals is hereby thankfully acknowledged.

Dr. J. Cheney of the Dept. of Veterinary Parasitology, Univ. of Nairobi, kindly identified intestinal parasites from faeces samples and C. H. S. Kabuye, and M. A. Hanid of the East African Herbarium helped in identifying grasses. Their assistance is greatly appreciated.

I wish to express my gratitude to Mr. C. A. W. Guggisberg who very kindly gave permission to use his extensive library and otherwise helped me with valuable advice. Mr. G. Adamson, Mrs. Bill Woodley and Mr. Pierre des Meules furnished me with information for which I am very grateful as well as for the kind hospitality offered by Mr. and Mrs. Adamson.

Dr. Hugh Lamprey allowed me to use the library of the Serengeti Research Institute and Dr. Hans Kruuk was most helpful with advice and encouragement; I am most grateful to both of them.

I want to express my deep appreciation to L. H. Bonnett and T. D. Morris, who read the manuscript and offered valuable suggestions for its improvement.

Finally, I want to thank the rangers of the Nairobi Park, whose help in tracking down the lions in the early mornings saved me innumerable hours of driving.

INDEX

Acacia drepanolobium 3, 23, 33, 84, 85,
    88, 104, 105, 107
  xantophloea 3, 107
  tortilis 107
  gerrardii 107
  seyal 107
  mellifera 107
  kirkii 37, 107
Acacia scrub or thorn scrub: *See*
    Ac. drepanolobium
Activity patterns 25, 27, 28
  daily 24
  daytime 24–29
  night 30–33
  synchronized 39, 44, 45
  displacement 40
Achyrantes aspera 107
Adaptation to killing large prey 89–90
  104
Ageing 6
Agonistic behaviour 42, 49
  aggression in cubs 75
  aggression during mating 66
  aggressive interactions 37, 50, 51,
    53, 54, 104
  threat displays 49, 50, 53
  defense and submission 51, 52
  redirected aggression 51
Alarm calls
  by mammals 100, 101
  by birds 101
Albizia isenbergiana 107
Amicable behaviour
  proximity 52, 53
  See also: Greeting and Grooming
Association of prides 16, 18·19, 60, 61
  of individuals 52, 53, 60, 61
  of male with various prides 19, 21,
    27, 30, 59
Athi Plains 1, 2

Baboon 3, 46, 80, 100, 101, 107, 112
Balanites aegyptiaca 3, 102, 107
Bear, brown 6, 112
  grizzly 7, 112
  polar 7, 112
Biomass of ungulates in NNP 77–79
Bothriochloa insculpta 107
Brachylaena hutchinsii 107
Breathing 38
Buffalo 3, 32, 80–82, 91, 100, 112
Bushbuck 107, 112

Carissa edulis 107
Caribou 112
Cat, domestic 30, 45, 53, 54, 63, 70,
    74, 85, 89, 97, 112
Cheetah 3, 88, 105, 107, 112
Chimpanzee 70, 71, 76, 112

Clawing tree 42, 45, 60
Climbing tree 37
Combretum gueinzii 107
Commiphora trothea 107
Communication. *See* Vocalization
    and Marking
Cordia ovalis 107
Cow, domestic (cattle) 46, 80, 112
Crane, crested 101, 112
Crocodile 3, 112
Croton megalocarpus 107
  dichogamus 107
Crow, pied 101, 112
Cubs
  development of 69, 74–76
  reaction to the male 74
  activity cycles of 28–29
  protection of 71–72
  suckling 24, 69, 70, 72
  mortality 68, 69, 105
  competition between 70
  abandoning of 68
Cymbopogon 107

Defecating: *See* Elimination
Dentition (of lion) 6, 7, 69
Description of lion 6–9
Digitaria macroblephara 107
Dik-dik 3, 107, 112
Diseases 38
Distances covered: *See* Walking
Distribution of lions 10
Dog, domestic 30, 34, 97
  wild (hunting dog) 3, 13, 44, 45,
    87, 97, 104, 112
Drinking 23, 24, 66, 85, 96–97
  times of 85, 96, 97
  duration of 97

Eagles 74, 112
Ear spots 7, 58
Eland 80, 95, 107, 112
Elimination
  urinating
  defecating
Elk, American 84, 112
Embakasi River 1–3, 77

Feeding 24, 29, 66, 91–96
  on vegetation 80
  dragging carcase (or carrying) 61,
    70, 91, 92
  position 92
  scraping near carcase 93, 94
  sequence 93–95
  of cubs 70, 71
  shaking of carcase 94
  consumption 94–96

Felidae 6, 9, 34, 40, 41, 46, 60, 73, 89,
    90, 91, 92, 94, 104, 110
Felinae 7, 29, 40, 73, 89, 92, 93, 104,
    105, 110
Flight distance 83, 100
Fox, bat-eared 3, 112

Game census 4, 77, 81
Gazelle, Grant's 53, 57, 78–83, 88,
    90, 100, 107, 112
Gazelle, Thompson's 78–83, 98, 100,
    107, 112
Genet 3, 112
Giraffe 82, 107, 112
Greeting (headrubbing) 25, 44, 48, 49,
    63, 76
    sessions 45
    description of 48–49
    development of, in cubs 74, 75
Grewia similis 107
Grooming 24, 25
    self 39–41, 45, 75, 76
    social (mutual) 39, 41, 46, 49
    parts groomed 40, 46, 71
    of cubs 71, 75
    sessions 39, 45
    function of 40, 44–47
    development of, in cubs 74, 75

Habitat
    preferences for resting 21–23
    preferences for hunting 23, 33
Hare 3, 95, 112
Hartebeest: See Kongoni
Headrubbing: See Greeting
Hippo 3, 112
Home ranges 15–19
    definition 15
    size of 18, 20, 32
    day and night ranges 15, 33
    utilization of 20, 33
    overlap 15–19
    changes during study 16–19
    relationship of, to each other 16–19
    core areas 20, 21
Horse, domestic 34, 112
Hunting 25, 28, 29, 42, 54, 66
    night 23, 85
    day 25, 28
    searching 30, 85
    killing 44, 57, 66, 89–91, 105
    muzzle-hold possible evolution of 44, 90
    preference ratios 82, 88
    direction of wind 87
    stalking 66, 74, 87, 88
    multiple killings 91
    killing frequency 95, 96
    success 97
    See also: Prey
Hunting dog: See Dog, wild
Hyaena, spotted 3, 12, 13, 15, 54, 68,
    88, 101, 105, 107, 112

Ibis, Hadada 101, 112
Identification method 5, 108, 109
Impala 82, 107, 112
Injuries 37–38, 61

Jackal 3, 101, 107, 112
Jaguar 7, 112
Juvenile spots 6

Kafue Park 12, 68, 82
Kanha, India 20, 30
Kenya National Parks 2
Killing: See Hunting
Kitengela Conservation area 2, 106
Kongoni 53, 57, 66, 78–84, 88–90, 92,
    93, 95, 100, 104, 107, 112
Kruger Park, S.A. 12, 20, 34, 68, 82,
    87, 106
Kudu 81, 112

Lantana viburnoides 107
Lake Manyara Park 11, 12, 19, 37
Leadership 54
Leopard 3, 7, 8, 68, 107, 112
Lippia javanica 107
Litters: See Reproduction
Locomotion: See Walking and Climbing

Mane, grooming of 40, 75, 76
    development of 75
    extent and colour of 7–9
Marking 36, 59
    with head 59, 60, 75
    urinating and scraping with hindfeet
        39, 60, 75
    scent spraying 59, 60, 63, 75
    reaction to 60
Mating behaviour 47, 62–67
    oestrus behaviour 63
    oestrus cycle 62, 63
    elements of 47, 64, 65
    frequency of copulation 66
    season 63
    aggression during 64–66
Mbagathi: See Embakasi
Meerkat 112
Melanism 8
Monkey, vervet 3, 46, 80, 107, 112
Moose 90, 97, 112
Muzzle-hold: See Killing under Hunting

Nairobi National Park 77, 85, 96, 102,
    103, 105, 106
    physiography 1–3
    history 1, 2
Ngong Hills 2
Ngorongoro Crater 9, 12, 30, 78

Olea africana 107
Ostrich 96, 112

Pantherinae 7, 29, 34, 40, 89, 110
Parasites
ecto 38
endo 38, 84
Pavonia patens 107
Pennisetum mezianum 107
Phyllanthus sepialis 107
Pig, domestic 70, 112
Play 25, 72, 74
cubs 25, 26, 28, 29, 73, 74
adults 25, 28, 73, 74
hours of 25, 26, 29
elements of 73
functions of 73
social 73
solitary 74
with objects 74
with carcase 54, 74
with prey 91
Plover, crowned 101, 112
Population (lion)
structure 11
composition 11
density 12, 105
changes in 11, 79, 103
Prey
selection of 80–85, 87, 88
predator ratio 78, 79
changes in, population 77–79
animals' reaction to lions 100
biomass 78, 79
Prides
composition 13, 14, 103
size 12, 14
number of 13, 15
relationships between 60, 61
Puku 82, 112
Puma 80, 84, 94, 112

Range: See Home Range
Rank order 53, 54
Reaction to cars 98, 99
humans 98, 99
weather 102
Recognition method: See Identification
method
Reedbuck 80, 90, 107, 112
Reproduction 103, 106
parturition 69
gestation period 67
sexual maturity 62
litter size 67, 68
sex ratios in litters 67, 68
See also: Cubs and mating behaviour
Resting
positions of rest 29, 30
hours of 25
points 20–23
Rhino 3, 100, 101, 107, 112
Rift Valley 1

Roaring 7, 55, 75
communal, in captive lions 45, 57
communal, in wild lions 44, 45, 57
function of 55, 58
reaction to, by cubs 58
Rolling 43

Scavenging 57, 88, 92
Schrebera alata 107
Scraping with hindfeet while urinating:
See Marking with front feet
near carcase 93, 94
Serengeti 9, 11, 30, 45, 81, 86, 88,
99, 105
Sex ratio in lion population 12
Sexual behaviour: See Reproduction
and Mating behaviour
Sexual dimorphism 54, 69, 75, 76
Spring hare 3, 112
Squirrel, ground 73
Steinbuck 80, 95, 107, 112
Stork, marabou 80, 101, 112
Stretching 41

Tail flicking or jerking 42, 43, 67
Taxonomy of lion 79, 110
Termitaria 3, 23
Territory 83, 103, 104, 106
Themeda triandra 107
Thomson's gazelle: See Gazelle,
Thomson's
Thorn scrub: See Acacia drepanolobium
Ticks 46, 112
Tiger 6–8, 10, 19, 20, 30, 54, 60, 70,
76, 80, 87, 89, 91–94, 97, 102,
105, 112

Urinating: See Elimination

Vibrissae 5, 108
Vocalization 55–57
cub-calling 72
mating roar 64
cubs' distress call 71, 72
adults' distress call (moaning) 58
See also: Roaring
Vultures 88, 102, 112

Walking 24
pattern of 35, 36
hours of 25, 27
speed calculations 34, 35
pacing 34
trotting 35
galloping 35
distances covered 30–33
on roads 34
Warthog 66, 81, 82, 89, 90, 92, 95, 97,
104, 107, 112

Waterbuck   81, 82, 107, 112
Weather: *See* Reaction to
Wildebeest   61, 71, 78, 79, 80–85,
      88–91, 96, 100, 104, 105, 107,
      112
Wolf   16, 44, 61, 64, 80, 84, 90, 97,
      104, 112

Yawning   44, 45

Zebra   46, 74, 78, 80–85, 88, 95, 97,
      100, 104, 107, 112
Ziziphus mucronata   107
Ziziphus abyssinica   107